Interface for an App—

The design rationale leading to an app that allows someone with Type 1 diabetes to self-manage their condition

Synthesis Lectures on Human-Centered Informatics

Editor

John M. Carroll, *Penn State University*

Human-Centered Informatics (HCI) is the intersection of the cultural, the social, the cognitive, and the aesthetic with computing and information technology. It encompasses a huge range of issues, theories, technologies, designs, tools, environments, and human experiences in knowledge work, recreation and leisure activity, teaching and learning, and the potpourri of everyday life. The series publishes state-of-the-art syntheses, case studies, and tutorials in key areas. It shares the focus of leading international conferences in HCI.

Interface for an App—The design rationale leading to an app that allows someone with Type 1 diabetes to self-manage their condition
Bob Spence
www.morganclaypool.com

ISBN: 9781636391557 print
ISBN: 9781636391564 ebook
ISBN: 9781636391571 hardcover

DOI 10.2200/S01098ED1V01Y202105HCI050

A Publication in the Morgan & Claypool Publishers series
SYNTHESIS LECTURES ON HUMAN-CENTERED INFORMATICS
Lecture #50
Series Editor: John M. Carroll, Penn State University

Series ISSN 1946-7680 Print 1946-7699 Electronic

Interface for an App—

The design rationale leading to an app that allows someone with Type 1 diabetes to self-manage their condition

Bob Spence
Department of Electrical and Electronic Engineering, Imperial College London

SYNTHESIS LECTURES ON HUMAN-CENTERED INFORMATICS #50

MORGAN & CLAYPOOL PUBLISHERS

ABSTRACT

This book is an account of how I addressed the need for a smartphone app that would allow someone with Type 1 diabetes to self-manage their condition.

Its presentation highlights the major features of the app's interface design. They include the selection of metaphors appropriate to a user's need to form a mental model of the app; the importance of visible context; the benefits of consistency; and considerations of a user's cognitive and perceptual abilities. The latter is a key feature of the book.

But the book is also about the design process, and especially about the valuable contributions made by the many focus group meetings in which design ideas were first presented to people with Type 1 diabetes. Their critique, and sometimes their rejection, of interface ideas were crucial to the development of the app.

I hope this book will prove useful for teaching and design guidance.

KEYWORDS

App UX design, user interaction, diabetes, app usability, mental model, perception and cognition, requirements

For

LEAH

and

JEREMY

Contents

Terminology

In the following pages we explain some of the terminology that may not be familiar to the reader. They include:

Affordance

Mental models

Metaphors

Dialogues

Exploration

I hope they are useful. I should stress that the rest of the book does not assume that these explanations have been read or understood.

Affordance

FAMILIARITY

Throughout this book the concept of "affordance" is all-pervading, so it is essential to clarify what the term means. Fortunately, this is an easy task, because it is a term that is used quite naturally in conversation. Here are three examples:

- "That door affords (i.e., allows, facilitates) entry to a café."

- "The app must afford (i.e., support) the choice of a favourite meal."

- "The next traffic circle affords access to two motorways."

Let's choose a real example, the door shown in Figure A. A glance at the door, and especially at the vertical plate on its right-hand side, suggests that the door affords opening by pushing on that plate. That glance, together with the assumption about the vertical plate, has identified the door's **perceived affordance**. If we push on that plate and the door opens, we discover its **actual affordance**.

Why do we introduce two definitions of affordance? **Because we want them to be identical**, and that is what a designer—in this case of doors—must try to achieve. Figure B shows another door, whose **perceived affordance** is that one must pull the door open. Only by testing will we know if that is also the **actual affordance**.

A situation that really will cause confusion is shown in Figure C.

Deploying an Affordance

To reach the café behind the door in Figure A, that door's affordance must be **deployed** (i.e., activated, engaged) by some action on the part of the user. For that particular door, the required action is signified by the vertical plate, which is quite reasonably called a **signifier**.

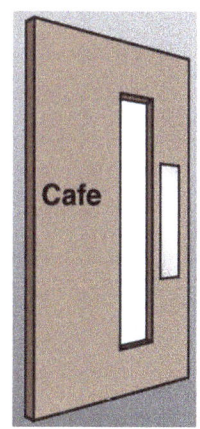

Figure A: This door apparently affords opening by pushing on the vertical plate.

Figure B: This door apparently affords opening by pulling on the handle.

There are many forms that a signifier can take: examples include the vertical plate seen in Figure A and, in the digital domain, familiar icons. In many cases, the design of those icons may not be easy, since their purpose is to ensure that they are correctly interpreted by a user.

Signifiers: Do we need them?

But a reasonable question that an interface designer might ask is "should an affordance be accompanied by a signifier?" That question is well illustrated by Figure D, part of a collection of photographs. That presentation may well offer more than one affordance, each deployed by a different user action: horizontal scroll, vertical scroll, mouse click, mouse double-click, continuous mouse-down, and so on. If a visible signifier were to be assigned to all those actions the display would become unacceptably crowded: and that may be the principal reason justifying their absence. Thus, the implicit assumption behind an absence of signifiers may be an interface designer's view that, through training, exploration, friendly advice, or other means, a user will quickly become familiar with the

Figure C: A very confusing situation!

various affordances and not need to interpret signifiers. Not an easy design decision!

Figure D: Many affordances but an absence of signifiers.

Mental Models and Metaphors

Much is written about mental models, but mostly one is left with the feeling that they are difficult to understand—ephemeral almost. Fortunately, nothing less than a concise definition has been provided by Jacob Neilsen of the Neilsen Norman consulting group:

> **Mental model:** what a user believes they know about a user interface.

Thus, a mental model is based upon **belief**, **not facts**. We also note that, as the user gains experience of an interface, their mental model will be **updated**, and often quite frequently. In other words, a mental model will continually be in a **state of flux**.

Familiarity

Since an app of any complexity will offer many possible interactions, the truly enormous challenge presented to a user-interface designer is now apparent. How can we help a user to form a belief—and preferably a very simple one—about how an app works? An answer is suggested by the term "familiarity". If the property of an interface can be related to something simple and familiar—and often physical—then that property of the interface may be more easily remembered. That "something" can be called a "**metaphor**".

Two Examples

Two illustrative examples of useful metaphors can be taken from the Chapter 4. The first is shown in Figure E: a strip of paper pulled back around two uprights so that all of the paper strip can still be seen. A digital embodiment of that metaphor is in fact used in Chapter 4 to provide a diary for the app user, simply by showing successive days on the "strip of paper". Once explained, not easily forgotten.

The second example—a horizontal stack of transparent vertical plates—is shown in Figure F. On each plate there is a rectangular non-transparent region so sized and positioned that, irrespective of which plate is at the front, no region will ever be totally hidden from view. A digital embodiment simply assigns four important personal attributes (Food, Exercise, etc.) to the four regions, and allows any plate, together with its region, to be brought to "the front" while maintaining the valuable context of the remaining regions.

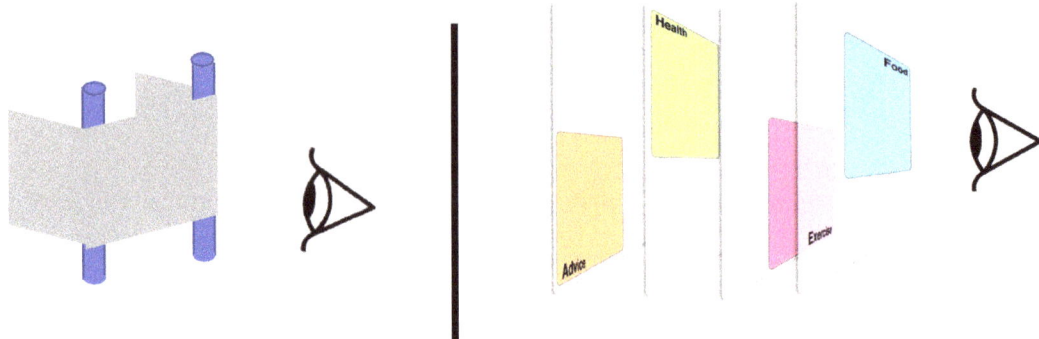

Figure E: A user can see all the paper strip, albeit with some parts distorted.

Figure F: A horizontal stack of transparent vertical plates, each containing a non-transparent colored region.

Dialogue

Conversation

When two people engage in a conversation, we say that a **dialogue** is taking place. That dialogue is usually quite complex, containing an unordered, frequently interrupted, and often uncompleted collection of questions, statements, answers, and opinions as well as utterances (e.g., Hmmm, Wow!) of many kinds together with a variety of facial and hand gestures. A documented record of most conversations would make little sense.

Interaction with a Computer

There is no place for such an unstructured dialog between human beings and computers. Nevertheless, even a carefully designed software system can benefit from a check to see if any features could cause a user to experience less than a smooth dialogue. Thus, while there is plenty of guidance regarding interface design, useful checks can instead draw the designer's attention to situations that must be avoided. Below is a list of some situations that should be checked: a more comprehensive list with more extensive discussion can be found in Elmqvist et al. (2011) and within this book (Chapters 5 and 13).

Checks

- Avoid situations in which a user might not notice a change in what has been displayed, a situation that is far more common than expected. Try identifying the difference between the two images at right, for example. A typical way of ensuring that Change Blindness[1] as it is called does not disrupt a dialogue is to ensure that most if not all visual changes are "fluid", ensured by animation of some kind. A duration of about 300 msec is suggested.

[1]   See Rensink et al. (1997).

- Ensure that a visual change happens after **every** interaction, employing a visual "shudder" if no change is permitted.

- Ensure that the location of a touch is identical to that of the corresponding visual change. Interactions should be integrated with the visual representation

- Ensure that interaction "never ends". There should be no "dead ends" beyond which interaction cannot proceed.

- To minimize navigational difficulties, provide visual context showing available transitions.

- Ensure that there is an immediate response to every user action.

Those are just some of the useful checks that can be made.

Exploration

Curiosity

Curiosity can take many different forms. From the simple "how much would it cost to borrow £1000?" to the far more challenging "if income tax is raised by 1% what effect, if any, will it have on unemployment?" These and similar questions imply that something might be **changed**, and it is the result of that change that is of interest.

Exploration

Some of the questions regarding change will have specific values in mind. Thus, X will be changed from 3 to 4—what will be the change in Y? In contrast, there are many situations where a user wants to say, "let's vary X and see what happens to Y". They simply want to **explore** the relation between X and Y.

There are two approaches to exploration. In one, the user simply chooses, often arbitrarily, a value of X, waits for Y to be calculated, then chooses a different value for X, and so on. The drawback of this approach to exploration is that the user will have difficulty in remembering the X,Y pairs that have been generated, especially if the underlying calculation takes more than a few seconds.

Dynamic Exploration

The second approach depends upon being able to perform the calculation of Y, given the value of X, in less than about 1 sec, allowing the user to vary X continuously while observing the corresponding values of Y. Such **dynamic exploration**, in which X is smoothly and manually varied, provides the user with a profoundly different experience: within a few seconds they can form a **mental model** of the relation between X and Y.

Predictions

The effectiveness of dynamic exploration requires the calculation of Y within less than a second after X has been specified. If, conventionally, that calculation cannot be carried out so quickly, the enormous benefit that accrues from dynamic exploration nevertheless serves as strong encouragement to find a faster means of calculation. Sometimes that will not be possible, and the benefit cannot be achieved. But in many cases, it can—read on!

Example

A good example of the benefit of dynamic exploration is provided by the app whose development is the subject of this book. A user with Type-1 diabetes is constantly aware that their blood glucose level (the Y of the calculation mentioned above) must be kept within safe levels, and that it is significantly affected by the carbohydrate value (X) of any meal they consume. Fortunately, for any carbohydrate value, the corresponding blood glucose variation over the next hour or so can be predicted with acceptable confidence. The user will be able, very quickly, to "home in" on a desired approximate range of carbohydrate values which can then be translated flexibly to a meal of interest.

Chapter Summaries

1   **Introduction** *This chapter describes the background of the project for which the app design described in this book was created.*

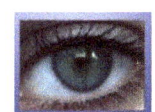

2   **Requirements** *In this chapter, I look at how we proceed from the original vision regarding an app to decisions about what activities of the intended user should be supported. A potentially useful and very simple concept in such a development is that of affordance.*

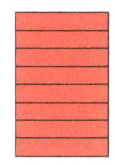

3   **Structure and layout** *The requirements identified in Chapter 2 are examined to identify any structure that would benefit design decisions. As a result, a proposal is made regarding the overall general appearance of the smartphone app.*

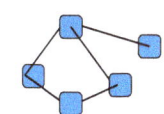

4   **Interface metaphors** *In this chapter we describe two metaphors on which the interface design is based. The main criterion for selecting those metaphors was their ability to help the user form an easily deployed mental model of the interface.*

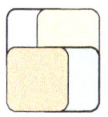

5   **Dialogue** *Use of the app will involve the user in several separate interactions, often by touch. While it is essential that each individual interaction should be designed with care, it is also necessary to consider the collection of interactions as a rich dialogue between user and app. Design considerations exist that relate to single interactions, and these are discussed with respect to the Diary and the "stack" of personal regions. Design issues associated with the human visual system and cognition are discussed.*

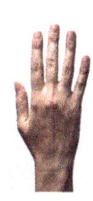

6   **Data entry** *The Food affordance comprises seven sub-affordances, one of which ("carb") allows a user to enter the carbohydrate value of a chosen meal and the time of its intended consumption. In this chapter we design the appropriate icon ("signifier") in the Food region that allows this activity to be initiated as well as the tool that is provided to allow those entries to be made.*

Food

Carbs

45g

7   **Exploration** It is extremely helpful if a user can manually and smoothly vary a carb value and immediately see the effect on predicted blood glucose level over the next hour or so, together with a recommended insulin dose. Such an activity is called "dynamic exploration" and is made possible by our ability, based on machine learning, to make such a prediction.

8   **Favourites** We are all creatures of habit, well illustrated by an individual's collection of favourite meals. A user should therefore have easy access to such a collection, and a subordinate affordance "Favourites" provides such access: that access is provided through the convenience of a personalised scrollable menu of favourite meals.

9   **Photograph** As use of the app proceeds it is anticipated that a user will wish to make additions to their collection of favourite meals. They must therefore be able to photograph such a meal and separately arrange for its appearance in the scrollable collection of Favourites. With consistency with other sub-affordances in mind, we design the tool for such a situation.

10   **Exercise** Exercise is one of the four personal affordances. Here we design the tool that permits the entry of Exercise parameters such as type (e.g., squash), extent (e.g., 30 min), scheduled time (e.g., 4PM), and aerobic level.

11   **Health** Health is one of the personal affordances. We design the appearance of its region and the tool that allows appropriate data to be entered. We also use the Health region to illustrate a simple way of determining what temporal data will be presented in the Diary, an approach that generalizes to all personal regions.

12   **Advice** One essential function of the app is to provide advice to the user. That advice may range from regular recommendations concerning insulin doses to alerts regarding blood glucose levels that require urgent attention and acknowledgement. Conventionally, alerts are often delivered by pop-up windows, but the rationale behind the approach adopted here is that the user's mental model would suffer from such an additional dialogue feature and that the expected location of advice is in the Advice region and its tool.

13   **Dialogue check** *The aim of an app is to support an effective and enjoyable dia-logue. Experience has taught us, the hard way, that many undesirable features are exhibited by apps that fail that test. In this chapter, a published set of design guidelines allow us to gain an impression of the quality of the app's design.*

14   **Conclusions** *One can only draw subjective conclusions about the usability of an app that, principally for virus considerations, has not been evaluated. Neverthe-less, for debate, I present four subjective claims.*

15   **Reflections** *In this brief chapter I reflect upon the concept of affordance and the way it has influenced the progress of design, and suggest how the affordance representation might beneficially be used in a physical design environment.*

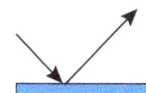

16   **Colleagues** *I did not work alone on the interface design: I drew inspira-tion and advice from a variety of people, some concerned directly with the project, some not. They are not included as coauthors for the simple reason that this book contains my own very personal views about interface design that I must not attribute to others. Brief biographies of the people who*

have influenced the app's design are included here as a way of saying "thanks for a wonderful experience."

CHAPTER 1

Introduction

This chapter describes the background of the project for which the app design described in this book was created.

1.1 AN OFFER

An offer I couldn't refuse triggered the story told by this book. One requirement of a project proposed by Imperial College London was the involvement of a person experienced in the field of Human–Computer Interaction (HCI), of which usability of the project's intended product was of primary concern. Having been around in that field for over 50 years I felt that I might be able to fill that role. Another attraction was the opportunity to develop an aid that could materially affect the lives of those many people suffering from diabetes and other chronic conditions. A third was the opportunity to draw upon the experience I had gained, over the past 50 years, in applying HCI principles to product development. And an undeniable attraction was the prospect of collaboration with clinicians, AI specialists, engineers, and others.

An invitation to join the project was enthusiastically accepted.

1.2 VISION

It was the vision of one of my colleagues, Dr. Pantelis Georgiou, that a person suffering from Type 1 diabetes would benefit enormously from a smartphone app that would help that person to self-manage their condition. The realization of that vision involved a collaboration between a team of people having expertise in the many relevant disciplines. My own contribution was to the design of the app's interface—broadly speaking, the dialogue taking place between the app and its user. The name of the project was ARISES. ARISES (Adaptive, Real-time, Intelligent System to Enhance Self-care of chronic disease) is a mobile platform (Spence et al., 2020) facilitating a decision support system for people with Type 1 diabetes (T1D) to improve the efficacy of current care and reduce the burden of managing diabetes in daily living.

1.3 TYPE-1 DIABETES

It really is no fun being a diabetic. It requires a constant awareness of your blood glucose level and the careful management of the factors that influence it—your lifestyle in general, though mainly your food consumption, exercise and health. The science behind Type 1 diabetes is complex, but if you're seeking a two-sentence summary, here it is:

> *Type 1 diabetes is a chronic condition characterized by insulin insufficiency occasioned by the autoimmune destruction of pancreatic beta cells. Subcutaneously administered replacement therapy is the mainstay treatment, and can be delivered as multiple daily injections or via a continuous subcutaneous insulin infusion pump.*

Of primary and continuous concern to the user is their blood glucose level, which must be kept within safe bounds.

1.4   PERSONAL

This book has a single author: me. At first that may seem at odds with the collaborative nature of the team that developed the smartphone app, and the extensive support the team provided me with during my involvement with the design of its interface. That is easily explained. Interface design, involving consideration of human cognition and perception, requires design decisions that are typically influenced by personal views and experience, notwithstanding the enormously valuable input received from colleagues and representative users. Therefore, I must not associate all the views and design decisions described in this book to my colleagues, especially since some of my proposals were controversial or just plain wrong (and are identified as such). My grateful thanks to my colleagues and others is made clear in the very brief biographies at the end of the book.

1.5   AFFORDANCES

A concept that may be unfamiliar to the reader is that of "affordance." I have found it to be helpful to my thinking about interfaces, but I have tried to ensure that its mention in the book is not intrusive. The concept is briefly explained in Chapter 2.

1.6   FOLLOW-UP

Where relevant to the text, references to the literature are provided.

CHAPTER 2

The Requirements

In this chapter, I look at how we proceed from the original vision regarding an app to decisions about what activities of the intended user should be supported. A potentially useful and very simple concept in such a development is that of affordance.

My first step in the creation of the app was very well defined. It required no experience of HCI, computer science, or programming. And I only had to communicate with two people (Figure 2.1). One was the person whose vision it was to enable people with Type 1 diabetes to self- manage their condition. The other was the medical lead on the project: he was and is both an experienced doctor and an expert on diabetes, with extensive clinical experience.

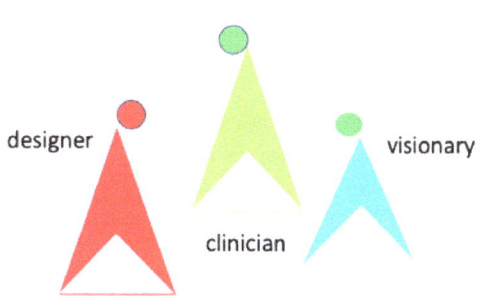

Figure 2.1: The actors involved in requirement specification.

What we discussed can be summed up in one, though perhaps unfamiliar, word—**affordance**. This term will have relevance throughout the book. For those who come new to the term the following very short section offers an explanation and introduces just two important terms: affordance and signifier.

2.1 AFFORDANCES

Affordance can briefly be illustrated by considering a door leading into a café (Figure 2.2). That door is said to **afford** (i.e., allow or facilitate) opening by pushing on the right-hand edge. The vertical plate extending over part of the right-hand edge is a **signifier**, hopefully allowing a user wishing to enter the café to correctly formulate, in their mind, a "perceived affordance." The affordance in this case is **deployed** (i.e., activated) by the user pushing the right-hand side of the door. (If the ver-

---

2   It is Gibson (1979) who coined the term "affordance" and is credited with first identifying the value of that concept. Norman's (1988) later discussion and illustration of the concept has ensured its wide recognition and application to interaction design.
Much has been written, some of it controversial, about affordance. See, for example, Victor Kaptelinin (2013), or the same author writing for the Interaction Design Foundation. In his book Where the Action is, Paul Dourish (2001) comments on affordance, as does Hinton (2015).

tical plate had been replaced by a handle the perceived affordance would probably have been that the door afforded opening by pulling). If actual and perceived affordances are identical, the signifier is well designed. With extensive use, a signifier is often not required, and is sometimes not provided on the assumption that a user will know, instinctively, by training or by experimentation, what affordance is offered: an example is the conventional scrolling action to review photographs.

2.2   AFFORDANCES NEEDED IN THE APP

At this initial stage input from the medical lead was obviously vital, with the app designer (me) on a steep learning curve.

Figure 2.2: A door that affords opening by pushing on the right-hand edge.

Extensive and continuous discussion allowed us to establish the affordances that must be associated with the app: in other words, the ways in which a person with Type 1 diabetes would use it. A clinical requirement was that the app must allow a user to enter details of an intended meal, their choice of planned exercise and their current state of health. That would allow recommended insulin doses to be calculated by the support system (Daniels et al., 2020) and presented for view. And an essential affordance must support useful views of temporal data such as current and past blood glucose events, insulin doses, and the state of the user's health.

Table 2.1 summarises the affordances that must be supported by the app, in each case with an illustrative statement to emphasise the relevance of the term "affordance."

Some of the affordances identified in Table 2.1 contain subordinate affordances. For example, seven subordinate affordances are associated with the Food affordance, as shown in the table. Four of them support entry—and viewing—of the carbohydrate content of a planned meal, its fat, and its protein content, and also drink. Another ("Explore") provides an opportunity for the user to manually explore the predicted effect, over the next hour or so, of carbohydrate content on blood glucose level. The user will also be afforded the opportunity to straightforwardly select a favourite meal ("Favourite") and to photograph a new one ("camera"). It is assumed that provision will be made for the user to view any entered data.

It bears repeating that Table 2.1 constitutes only the performance requirements of the app. At this stage the interface designer must ignore all considerations of the technology that will enable those affordances to be provided, as well as aspects of human cognition and perception that significantly influence the way they are to be deployed.

Table 2.1: Affordances that must be provided by the app

Food: the app must afford:

    entry of carb, protein, fat, and drink values and intended time of consumption

    choice of a favourite meal

    exploration of the relation between carb value, predicted blood glucose level and needed insulin doses

    the photographing of a new favourite meal

Exercise: the app must afford:

    the choice of exercise type, value, duration, and intensity

Health: the app must afford:

    entry and viewing of the user's state of health

Advice: the app must afford:

    a user's response to alerts and their acceptance or otherwise of recommended insulin dose

Diary: the app must afford:

    a user's examination and interpretation of recorded temporal data

CHAPTER 3

Structure and Layout

The requirements identified in Chapter 2 are examined to identify any structure that would benefit design decisions. As a result a proposal is made regarding the overall general appearance of the smartphone app.

3.1   REQUIREMENTS

Table 2.1, repeated here as Table 3.1, lists the affordances that the app must make available to a user. To identify useful groupings of those affordances we recognise that control of diabetes is largely *and jointly* influenced by four aspects of personal data: **food** consumption, **exercise** taken, the state of a person's **health**, and **treatment** administered. The user must be able not only to enter such personal data but also review it.

Table 3.1: A repeat of Table 2.1 which listed the affordances that must be provided by the app
Food: the app must afford:
entry of carb, protein, fat, and drink values and intended time of consumption
choice of a favourite meal
exploration of the relation between carb value, predicted blood glucose level and needed insulin doses
the photographing of a new favourite meal
Exercise: the app must afford:
the choice of exercise type, value, duration, and intensity
Health: the app must afford:
entry and viewing of the user's state of health
Advice: the app must afford:
a user's response to alerts and their acceptance or otherwise of recommended insulin dose
Diary: the app must afford:
a user's examination and interpretation of recorded temporal data

Independently, and at any time, a user must be able to examine a wide variety of **temporal** data, but especially their blood glucose level which is of major and continuous concern.

As a result of these considerations, we made a tentative decision—later accepted—regarding the location of affordances within the smartphone display (Figure 3.1). This simple figure emphasises that we are *not* dealing with a collection of disparate affordances that might typically be represented, *separately* and neatly arranged, on the screen of an iPad. Rather, during our design of the app, we must recognise that the wellbeing of a user is determined by the *totality* of the personal data completely independently of any temporal data.

3.2   STRUCTURE

To proceed further we summarise our approach by constructing a representation (Figure 3.2) of the five principal affordances. The combined clinical relevance of the four personal affordances is emphasised by their location on a black ring whose detailed implication is explained in Chapter 4. I have shown the temporal affordance separately, because the user might want to view historical data at any time and a facility for doing so should always be available.

It is essential to note that the affordances we see in Figure 3.2 are **not** visible icons: the representation shown in that figure has nothing to do *directly* with what the user sees on their smartphone or the **actions** they take during its use. The **appearance** of the app will only begin to be chosen in the next chapter, and detailed **interactions** in the chapter that then follows.

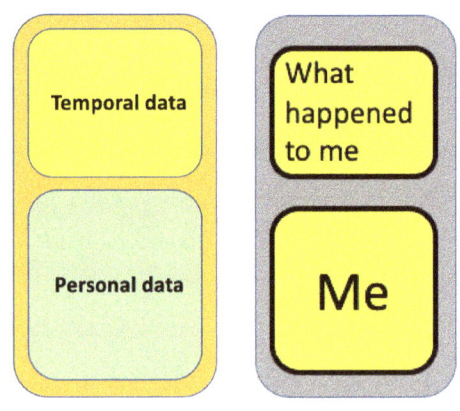

Figure 3.1: (left) Suggested layout of affordances on the smartphone display; (right) an informal summary that may be helpful for the new user.

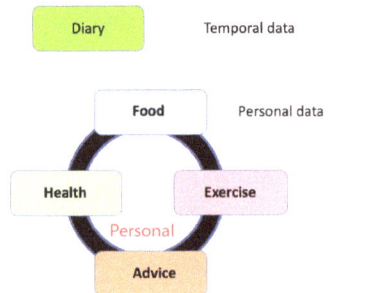

Figure 3.2: A first affordance representation.

CHAPTER 4

Interface Metaphors

In this chapter I describe two metaphors on which the interface design is based. The main criteria for selecting those metaphors was their ability to help the user form an easily deployed mental model of the interface.

4.1 USABILITY

Having identified, in Chapter 2, the affordances (Figure 4.1) and their grouping that the app is required to provide, our constant concern will be with the app's ultimate **usability**. Indeed, every design decision we make will automatically be accompanied by the question "how will this affect usability?" We don't know, of course, and even after a usability study with participant users we shall usually learn about overall performance, and not whether detailed option A was really better than option B.

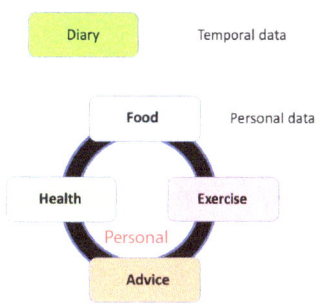

Figure 4.1: A first affordance representation.

4.2 MENTAL MODEL

The extent to which an app is usable will largely be determined by the ease with which a user can form and exploit a **mental model**. As Jakob Nielsen (2010) reminds us, "A mental model is what the user believes about the system" and that "What users believe they know about a UI [user interface] strongly impacts how they use it." Thus, "a mental model is based on belief, not facts." Also, that model is typically in flux: it will be formed by training and/or exploration, it will be modified by use and on occasion it may undergo a major rearrangement after a particular feature of the app is suddenly understood more clearly.

We have all had those experiences. Which leaves the designer with a challenge at this early stage of an app's design: how can we help the user to readily acquire a mental model of the app?

4.3 METAPHORS

I decided to see if metaphors could come to the rescue: one to handle temporal data and the other personal data. I'll look at those in turn, beginning with temporal data.

4.4   TEMPORAL DATA

In the pre-digital age, temporal data was usually handled by a diary. But that approach, for many reasons, does not carry over at all well to the technology we now have at our disposal and which we can exploit to advantage.

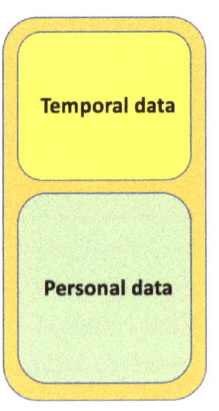

The display area available (see sketch of Figure 4.2) may be adequate for the presentation of a single day's record of blood glucose level, and for scrolling to allow similar views of adjacent days. But in addition to such *quantitative* data, we must also be able to draw, to a user's attention, *qualitative* data about significant episodes over a much longer period. Examples include the days on which glucose level reached dangerous hypo- and hyper-levels and the values of recommended insulin doses. How can we support such a variety of temporal interests, often stretching over many days? One answer is by a metaphor proposed in 1982.

Figure 4.2: Areas available for the two classes of data.

4.4.1   THE BIFOCAL DISPLAY

We are faced with the age-old challenge of "too much information, too little (display) area." The solution I proposed was published as long ago as 1982 by a colleague and myself (Spence and Apperley, 1982), and is best explained by reference to the metaphor we proposed.

Day 11	Day 12	Day 13	Day 14	Day 15	Day 16	Day 17	Day 18	Day 19	Day 20	Day 21	Day 22

Figure 4.3: Part of a diary.

Figure 4.3 shows a strip of days that form part of a diary. It is obviously unsuitable in that form for presentation on a smartphone display. The solution we proposed was based on the metaphor shown in Figure 4.4, where the long strip of pages, one for each day, is taken from Figure 4.3 and pulled back around two uprights though, importantly, at an angle that permits a sideways view of many adjacent days. A user's view of that arrangement could then be enlarged to occupy a conventional screen (Figure 4.5) and allow detailed viewing and interaction with a specific day's content—here, Day 15.

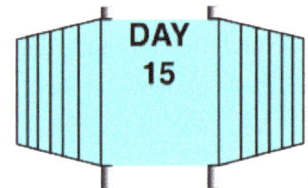

Figure 4.4: The distorted view of the diary pages of Figure 4.3.

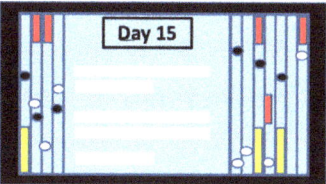

Figure 4.5: The bifocal view of the diary pages of Figure 4.3.

But at first it might seem that the adjacent, distorted days would be worthless, particularly since no text for those days would be readable in the narrow space. On the contrary, that small space can be put to excellent use, by providing a *qualitative* summary of events over a number of days. Have another look now at Figure 4.5, which is a view of the bifocal diary without the perspective introduced for purposes of explanation in Figure 4.4. For our application, for example, the red and yellow bars in the left and right distorted regions might indicate the proportion of each day that a patient's blood glucose was in the dangerous hyper- and hypo-glycaemic states, and black dots could represent insulin doses. If, from a qualitative sight of these adjacent days, a particular day appears to be of interest, it can be brought, enlarged, to the "central" ("Focal") readable region for detailed examination, simply by a scrolling action. The name "Bifocal Display" was assigned to the metaphor just illustrated.

The decision was taken to use the Bifocal Display concept to provide the user with a diary in the upper part of the smart-phone display as indicated in Figure 4.2. Figure 4.6, for example, shows today's recorded blood glucose level of a person with Type 1 diabetes from midnight until the current time indicated by the vertical dashed line. The term "Bifocal Diary"—henceforth referred to simply as the Diary—was suggested for the proposed interface for viewing and interacting with temporal data.

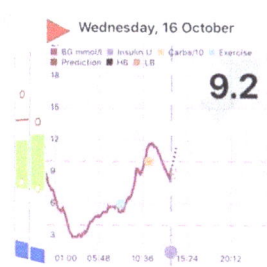

Figure 4.6: An early experimental design of the bifocal diary.

4.5   PERSONAL DATA

We must also propose an interface suited to a user's interaction with personal data, tentatively located in the lower half (Figure 4.2) of the smartphone display.

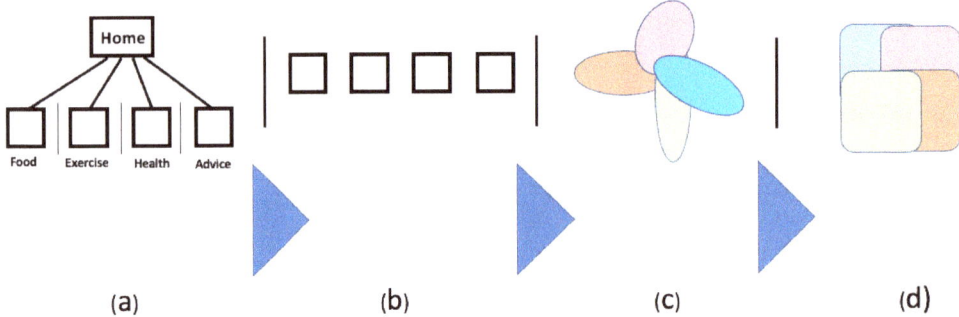

(a)   (b)   (c)   (d)

Figure 4.7: Development of the metaphor: (a) a conventional, but unsuitable, hierarchy; (b) the important, personal, affordances; (c) they are collectively important to wellbeing; and (d) a more memorable and space efficient presentation.

I arrived at a suggestion for handling personal data in a sequence of steps. Almost without thinking, I first sketched the familiar and conventional representation of a hierarchy of data (Figure 4.7a). My immediate dissatisfaction with that representation was threefold. First, I have very strongly held views about the danger that hierarchy poses to navigation, to the detriment of usability. Second, the implied independence of the four sets of personal data does not accord with the fact that it is their *combination* that defines the user's wellbeing. Third, I would want to be convinced about the need to include the concept of a "home state", especially because of its implication regarding hierarchy. I therefore removed all detail except for the four components of personal data (Figure 4.7b). At that point I recalled the advice I give to students that they might like to consider shapes other than rectangles when thinking about interfaces, and modified the representation to show (Figure 4.7c) a collection of four blobs but *overlapped* to emphasise their collective influence on wellbeing. Finally, conscious of the need to make the most of limited available display area as well as the effective use of each component of personal data, I then rearranged the four components as shown in Figure 4.7d, redrawn for convenience in Figure 4.8.

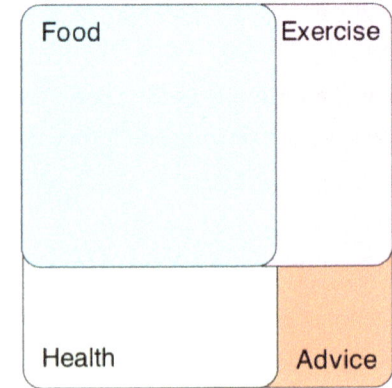

Figure 4.8: The four overlapping personal regions.

All that now remains to complete the metaphor is to assign each of the square regions to separate vertical transparent layers of a stack, as shown in Figure 4.9. The metaphor's relevance to our need is now apparent: a "frontal" view of that stack is identical with Figure

4.8, so I suggested that the stack metaphor leads to an easily recalled mental model. The necessarily two-dimensional display of Figure 4.8 is mapped, in the user's mind, into an easily recalled 3D mental model, with expected benefits for a user's interaction with personal data. As we shall see, obtaining a full view of the Exercise region (Figure 4.9) can be thought of as bringing the corresponding vertical transparent "plate" to the front of the stack, as it will henceforth be referred to.

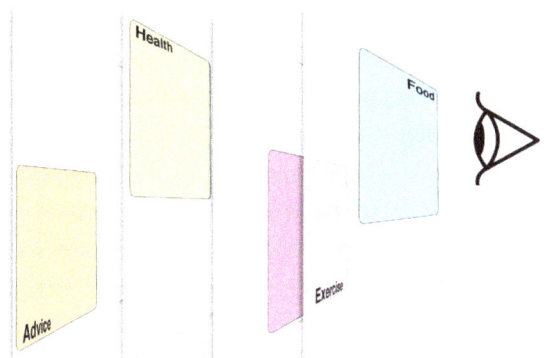

Figure 4.9: A horizontal stack of transparent vertical plates, each containing a "personal" region.

4.5.1   POPULATION OF THE PERSONAL REGIONS

The metaphor illustrated in Figure 4.9 supports the user's view (Figure 4.8) of the four regions. Those regions, shown blank in Figures 4.8 and 4.11, can now be populated with icons ("signifiers") by means of which the user can access the tools needed for data entry. Some idea of how those four regions will eventually appear can be gained from the "look ahead" provided in Figure 4.10. Detailed decisions about the population of the four regions are explained in Chapters 6, 7, 8, and 9 (for Food) and Chapters 10 (Exercise), 11 (Health), and 12 (Advice).

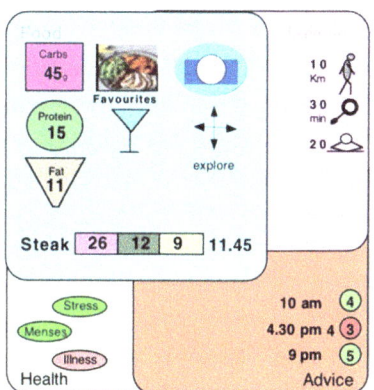

Figure 4.10: A glimpse of how the regions will be populated to allow access to tools for data entry.

4.6   DESIGN CONSIDERATIONS

It was decided to adopt both metaphors—the Bifocal Diary and the stack of Personal regions—as the basis for our interface design, a decision that has remained unchanged. We now examine some consequences of that decision.

4.6.1   VISIBLE CONTEXT

The presentation of Figure 4.10, based on the metaphor of Figure 4.9, serves a very important purpose: to provide visible context. For example, while entering data into the Food region (anticipated in Figure 4.10), the user can be aware of some existing choices of Exercise, Health, and Advice that can be seen in the "always visible" areas of those three partially overlapped regions. Irrespective of the "horizontal ordering" of the regions, a "full" view of a particular region is always accompanied by partial views of the other three regions. In other words, whichever region was the focus of attention, its context would be partially visible. The importance of context will be stressed in other chapters, and is widely discussed in the literature (Hinton, 2015; Dourish, 2001).

4.6.2   NAVIGATION

Another benefit of visible context has to do with navigation and a user's mental model of the metaphors. If, while food details are being viewed—with the Food region necessarily at the "front of the stack" (Figure 4.10)—there is a reminder to the user what other regions are immediately accessible and capable of being brought to "the front". My opinion is that navigational difficulties will be minimised in this way, partly because, as pointed out above, the metaphor of a stack may well provide a mental model that is easy both to acquire and recall.

4.6.3   A MISUNDERSTANDING

Unfortunately, the metaphor of the stack, and the appearance of Figure 4.10, initially led to a serious misunderstanding: it was remarked that "the 'always visible' part of each region is merely a 'tag'". Certainly not—that interpretation completely misses the anticipated advantage of visible context, notwithstanding the property—discussed with other aspects of interaction in the next chapter—that a touch on an "always visible region" will bring that region to "the front."

4.6.4   DESIGN FLEXIBILITY

Flexibility is always welcomed by a designer. It should therefore be noted that the size of each of the four regions can be independently chosen, and additionally be made to expand when moved to "the front". This freedom may help, for example, to accommodate more icons within selected regions, although there would be correspondingly less "always visible" space to display contextual data in the remaining regions. Such trade-offs are not unfamiliar to designers.

4.6.5   HOME STATE

I earlier discarded the idea of a "home state." As use of the app proceeds, the "horizontal" ordering of the four regions in the stack will change, and not necessarily remain in the same order after use. The term "nomadic" state therefore came to mind, though it was not received with enthusiasm. Later, the idea of a "Launch state" was mooted, but discarded because the user might well expect the last view of the interface to be the default opening view. I thought it might be best not to debate the concept of a "home state" any further.

4.6.6   VISUAL DESIGN

I often have to be patient when colleagues refer to interface design as "making things pretty". They might, for example, look at Figure 4.10 and wonder why I didn't use simple rectangles for region shapes rather than areas with rounded corners. The answer is available by glancing (I intentionally use that term) at Figure 4.11 and answering the question posed in its caption.[3]

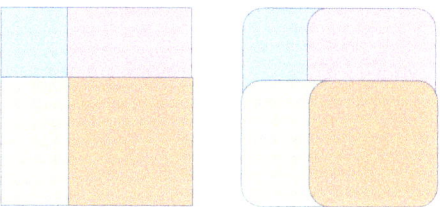

Figure 4.11: Which of these two images more readily suggests four overlapping areas?

---

[3]   I often refer to a design proposal as "tentative". That is because many decisions were only confirmed—at least temporarily—following their exposure, at regular meetings, to a group of representative users, usually around ten in number. Proposals were put to them and robust discussion followed, in meetings chaired by the clinical lead. Notes were made, and afterward informed the confirmation—or in some cases described later the discarding—of proposals. There is always a danger of a designer being blind to the defects of anything that they have personally invented, and I shall provide two examples later where feedback—from very articulate users—has led to proposals being abandoned, and for good reason.

CHAPTER 5

Dialogue

Use of the app will involve the user in a number of separate interactions, often by touch. While it is essential that each individual interaction should be designed with care, it is also necessary to consider the collection of interactions as a rich dialogue between user and app. Design considerations exist that relate to single interactions, and these are discussed with respect to the Diary and the Stack of personal regions. Design issues associated with the human visual system and cognition are discussed.

The previous chapter showed how the app's interface can usefully exploit two metaphors: the bifocal diary to support examination of temporal data and the "stack of regions" to allow the entry and viewing of personal data. Although we carried out simple checks to ensure that the interactions required of a user were reasonable, we now need to look into that issue more broadly. In fact, we must now recognise that we need to consider a profound aspect of the app and one which significantly affects usability, namely the **dialogue** that is taking place between user and app.

5.1 THE DIARY

The diary's interface metaphor (Figure 5.1), together with experience gained since it was invented, strongly suggests that the diary content should be scrollable. In one sense, that is all that needs to be said about interaction with the Diary. However, it is convenient here to mention that the bifocal view can usefully be complemented by what we have called a History view, suggested by Mark Apperley. It is, in fact, the more familiar view of time-varying data for most users. It is shown in Figure 5.2. It simply offers a conventional day-by-day plot of selected variables. Clinical advice suggested that a user should be offered a 7-day, 14-day, and 21-day view, though this format can easily be changed: Figure 5.2 shows a 14-day view, with "Today" in the right-most column.

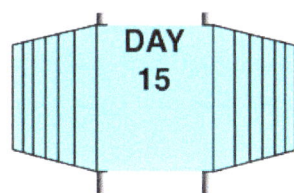

Figure 5.1: The metaphor underlying the Bifocal Diary.

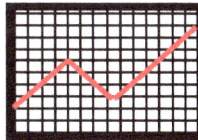

Figure 5.2: Appearance of the History mode of the Diary.

5.1.1 SWIPING AND SCROLLING: DO WE NEED SIGNIFIERS?

If the user is to be offered a choice of either the Bifocal or History mode of the Diary, we need to think about whether a signifier is needed. For the door used for illustration in Chapter 2 the signifier was a flat vertical strip on the side of the door that should be pushed. The question, therefore, is what signifier—if any—is needed to indicate that the bifocal display should be scrolled? My recommendation was "none". There are many examples of interactive devices where a signifier is not provided—the collection of images on a smartphone is one. And a benefit of having no signifier is that the interface does not get cluttered (an initial implementation of the Diary, before the History mode was added, is illustrated in Figure 4.6 and contained one signifier, the red arrow).

Is the same conclusion appropriate for the History mode? My view was that transitions between the Bifocal and History modes, and the extension and reduction of range in the latter mode, do not need signifiers, provided the actions required to cause those transitions are sufficiently simple and training is given. My proposal is illustrated in Figure 5.3.

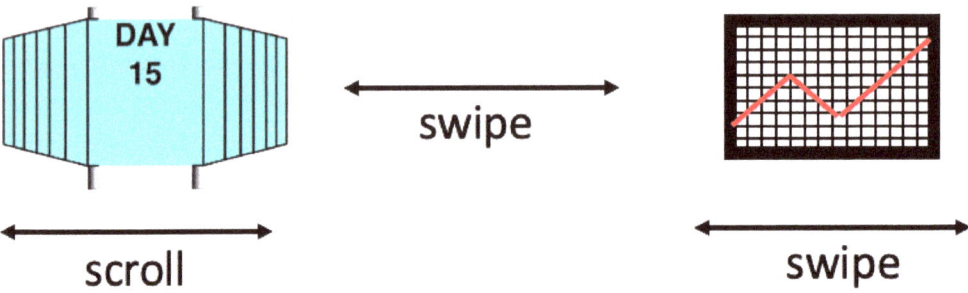

Figure 5.3: A right swipe (centre of figure) will change the diary from Bifocal to the seven day range of the History mode, and the reverse is true for a swipe to the left. Once in the History mode swipes will extend or reduce the range by 7 days.

It only remains to point out that scrolling of the bifocal mode might profitably be of the "snap-to" type, though user feedback and clinical recommendation might advise against. The interaction needed to allow the user to decide what variables are to be plotted in the history mode will be considered later in Chapter 11.

5.2   THE PERSONAL REGIONS

As for the Diary, the issue of interaction and related signifiers must be addressed for the personal affordances. As established in Chapter 4, the four personal regions use the metaphor of a stack viewed from the front, as illustrated in Figure 4.10 and repeated here as Figure 5.4. If, for whatever purpose, we are to make use of the full area of a region, then that region should be in full view and positioned at the front of the stack. Thus, for a region that is currently not at the front, some interaction is needed to put it there. The most obvious interaction is a touch anywhere on the visible part of the region of interest. I suggested that, for ease of recall, such an interaction should leave the other three regions in the order they were in before the interaction.

In making the above design decisions, of course, we have automatically identified suitable signifiers, namely, the visible parts of the partially masked regions. It can correctly be observed that the visible areas of the partially masked regions are acting as conventional tags. However, as pointed out elsewhere, those unmasked areas provide immensely valuable locations for visible context.

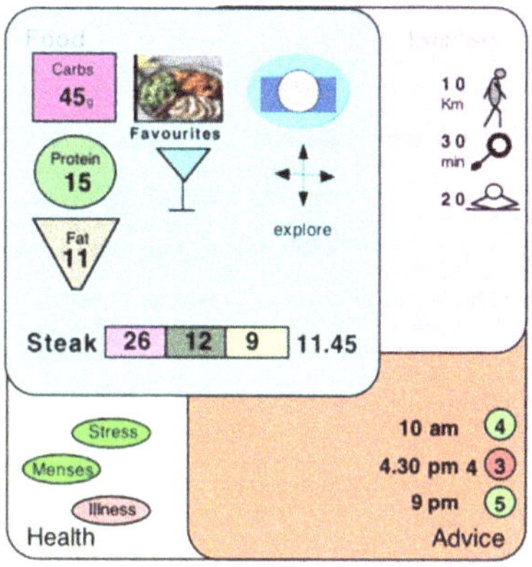

Figure 5.4: View of the Stack of personal regions (an initial design).

5.3   THE DIALOGUE

Designing for interaction involves much more than a decision about what happens in response to a single touch. Far from it. A primary influence on the usability of an app is the nature of the dialogue that takes place between user and app. Just as a conventional dialogue between two human beings can be rich, rewarding, complex, and characterized by far more than single word utterances, the success of any dialogue between a user and an app depends on more than single responses to touch. It is the rich combination and dynamic properties of touch, visual response, and timing that has an enormous effect on the usability of an app.

Valuable guidance regarding the design of a dialogue such as the one we are addressing has been provided by Elmqvist et al. (2011) in the form of eight guidelines for interactive systems. Some relate to single interactions, while others relate to the entire app. We have not yet designed the app's complete interface, so here we direct our attention to the guidelines relevant to what we have designed so far and return to the entire collection of eight guidelines when the final app design is established. We shall use the same headers (DG1 to DG8) as used by Elmqvist et al. (2011).

5.3.1   DG1 FLUID RESPONSE

Many people are surprised to learn that they might not notice an immediate change in an image, even if that change is substantial and, with hindsight, glaringly obvious. The effect is called Change Blindness. It was established a long time ago (Rensink et al., 1997; Simons and Lewin, 1997; Findlay and Gilchrist, 2003). If you find the effect intuitively difficult to believe then please go to Healey (2007): you'll be surprised at your inability to identify (afterward obvious) substantial differences between two images. To avoid Change Blindness, we require most visual changes to be "fluid". That means that the visible response to an interaction (typically a touch) must be animated with a duration of about 250 ms: some experimentation and judgement may be required during implementation.

There is another reason to employ fluid transitions, again well-documented (Robertson et al., 1991). It is based on the fact that, during an interactive dialogue, the user is forming a mental model of the interaction scheme. If a visual presentation changes instantly it is far more difficult for the user to modify their mental model than if the change had been fluid. Anyone who has doubts about the effect I have mentioned should watch the short but impressive movie associated with the paper by Robertson et al. (1991) and narrated by Stu Card.

5.3.2   DG2 IMMEDIATE VISUAL FEEDBACK

This subtitle speaks for itself. If the user performs an action (a touch, say, on an icon) and there is no discernible response, that user will be unsure about what has transpired. As Elmqvist et al. (2011) emphasise, this requirement applies to every interaction, not just "major" ones. We can, in fact, identify two situations where this guideline is especially relevant. One is when a user touches an inappropriate area of a region already at the front of the Stack. Another is when a scrolling action is attempted when the diary is in History mode, and responsive only to swipes. In such cases, and with all similar situations, the conventional "visual shudder" should occur. Otherwise, a check on our design so far confirms that *there can be an immediate visual response to every user action.*

5.3.3   DG3 "MINIMISE INDIRECTION IN THE INTERFACE"

This guidance suggests that, where possible, direct manipulation should be used so that interactions are integrated with the visual representation. This requirement is satisfied so far by our app. For example, a touch on a partially visible region causes that region to "move to the front", and the bifocal diary will respond immediately to a scrolling action.

We have checked that the design decisions made so far comply with guidelines DG1, DG2, and DG3. Remaining guidelines will be visited, in Chapter 13 when the app design has been finalized.

CHAPTER   6

Data Entry

The Food affordance comprises seven sub-affordances, one of which ('carb') allows a user to enter the carbohydrate value of a chosen meal and the time of its intended consumption. In this chapter we design the appropriate icon ('signifier') in the Food region that allows this activity to be initiated as well as the tool that is provided to allow those entries to be made.

6.1   CARBOHYDRATE VALUE ENTRY

In Chapter 4, we chose a metaphor—the four stacked regions—suited to a user's engagement with the four Personal affordances (Food, Exercise, Health, and Advice) (see Table 2.1). To address the issue of data entry we focus, in this chapter, on the "carb" sub- affordance that allows a user to enter the carbohydrate value of a planned meal and the time of its intended consumption.

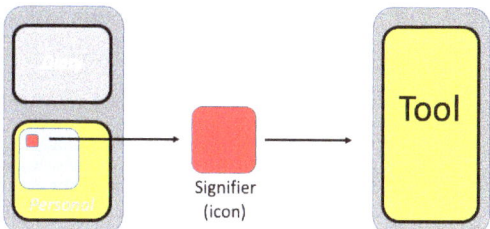

Figure 6.1: The scheme adopted to support the deployment of all personal affordances, annotated to illustrate the entry of carb and time values.

Before proceeding, I must anticipate a possible remark by the reader to the effect that "we've been entering numerical values for decades, so what's new?" The answer is "more than you would probably think". All I can say is "read on", because the development in this chapter will illustrate a remarkable consistency that emerged from a study of *all* the personal affordances, and it is my opinion that consistency will beneficially affect the user's mental model and hence the app's usability. The interface structure we shall choose in this chapter to facilitate data entry is precisely the same as that which supports *all* the personal affordances. The structure is illustrated in Figure 6.1 and annotated to facilitate our discussion of carb entry.

6.1.1   REGIONS, SIGNIFIERS, AND TOOLS

On the left of Figure 6.1 is a sketch of the whole of the smartphone's display area showing, in yellow, the location of the personal regions; the Food region is coloured blue. Within that Food region a red square is used to illustrate the approximate size and position of the "carb icon" that will allow the user to make the required data entry. An early illustration of finer detail of the Food region can be seen in Figure 6.2.

The area available for the Food region, as indicated to the left of Figure 6.1, suggests that it should only serve two principal purposes. One is to allow the user to view detail, in this case carb value. The other is to allow the user to locate and subsequently touch an icon that will, in response, provide a tool to be used for data entry. But that tool is too complex to be included within the Food region, and is best handled by allowing it to take over the entire display area (even including the area assigned to the Diary), as shown on the right of Figure 6.1. The user action that is needed to replace the content shown on the left with the detailed tool on the right is a touch on the carb icon in the Food region. This transition is suggested by the enlarged carb icon in the

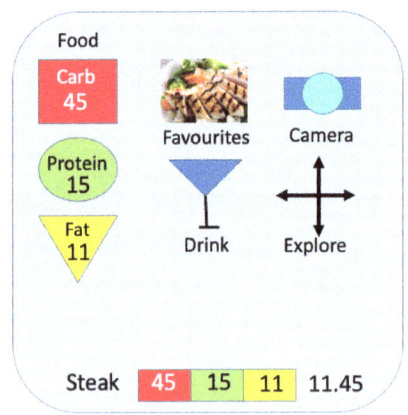

Figure 6.2: Appearance of the current version of the Food region.

centre of Figure 6.1. For completeness, we indicate that the reverse transition is achieved by a touch on an OK button appearing within the tool. Using the terminology associated with affordances we refer to the carb icon as a *signifier*. In what follows, we undertake a design that allows the entry of a carb value and the intended time of consumption.

6.2   DETAILED DESIGN

6.2.1   THE FOOD REGION

Our design of an affordance (e.g., carb) that is subordinate to Food will assume that, to deploy it, the Food region has been brought to the front of the stack. An illustration of the region's proposed appearance is shown in Figure 6.2. A signifier for each sub-affordance, of which there are seven, is provided by an icon having shape, colour, text, and position that satisfy a variety of requirements discussed below. A separate horizontal bar shows details of the last meal entered: name (optional), followed by carb, protein and fat levels, and time of consumption. In this chapter we confine our attention to the carb icon (Figure 6.3) located top-left in the Food region.

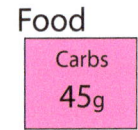

Figure 6.3: The carb icon in the Food region.

6.2.2   THE SIGNIFIER FOR CARB ENTRY

The colour and shape of the carb icon was chosen to distinguish it as far as possible from the other six icons: inset text (Figure 6.3) allows the user to be aware of the latest assigned carb value (in grams). Within the available area of the Food region the carb icon (like the others) was made as large as possible to allow for easy recognition at a quick glance as well as accurate touch.

6.2.3   THE TOOL

What follows in this chapter will identify several specific design considerations put forward with the confident expectation that they will individually and collectively enhance the usability of the app. Those proposals generalise to many of the affordances within the app.

With reasonable accuracy, a user will often be able to estimate the carb value of a meal they are planning, partly from experience and partly from recall from a well-known reference book (Cheyette and Balolia, 2016). For ease of use, I suggested that the entry of all carb values, as well as the intended time of consumption, would be entered by touch, (not by slider or scroller), using a very large keypad (Figure 6.4) like that typically available on a smartphone for making telephone calls.

The design rationale so far is simple. First, the ages of people with Type-1 diabetes span a very large range, and not all of them will be in perfect health. Therefore, design must take account of users with various forms of visual impairment and, perhaps additionally, with a finger/hand tremor of some kind. It can also easily be forgotten that many users will not be computer-literate: they must be catered to. So, the first design decision was to say that a single touch on the carb icon (top-left of Figure 6.1) would lead to the appearance of a tool offering a **numeric keypad** occupying most of the available screen area (Figure 6.4).

Figure 6.4: The tool contains a numeric keypad supporting the entry of carb and time.

Fortunately, without requiring a separate tool, and simply by including a single "clock" icon (Figure 6.4), we can allow the user to seamlessly enter the intended time of carb consumption in addition to its value. A click on the bistable clock face will immediately (1) highlight the clockface and (2) place the characters T.I.M.E (or "Now") in the results window. Entry of the time of consumption then proceeds as for carb value.

In addition to the design issues just addressed, some other issues need consideration, many relating to the perceptual and cognitive abilities of the human user.

6.3   THE HUMAN USER

6.3.1   CHANGE BLINDNESS AND THE NEED FOR ANIMATION

The phenomenon of Change Blindness and the need for animation of transitions were discussed in Section 5.3.1 of Chapter 5. They apply to the appearance and disappearance of the tool of Figure

6.3 following activation, respectively, of the carb icon within the Food region and the OK button located within the tool.

6.3.2   INCONSPICUOUS CONTEXT

A second consideration is associated with the benefit of visible context, mentioned in Chapter 4, Section 4.6.1. To exploit that benefit I chose (see Figure 6.4) to set the individual components of the keypad against a background which is a defocused version of the view the user has just left, rather than either a blank background (which could trigger the response "where did that go?") or the potential distraction of the actual or dimmed but detailed presentation of the view they have just left. I refer to this background as providing **inconspicuous visible context**. I'm essentially saying, "while the user's attention is focused for a few seconds on value entry, let's not offer any distraction. Note also that in providing some context—albeit inconspicuous—I'm trying to minimise the navigational hazard often associated with apparent hierarchy.

6.3.3   COGNITION

For a third consideration, look at a conventional physical keypad. The buttons are all the same size and arranged on a grid, but purely for cost considerations which do not (should not!) apply to a digital interface. Thus, by setting the Delete button in Figure 6.4 noticeably apart from the numeric buttons, and using a distinctive colour, I'm trying to help a user be aware that two functional groupings are present.

6.3.4   MEMORY

Fourth, to support the user, we arrange that the characters C.A.R.B appear first in the window, before any digits are entered. Is that really necessary? Yes, it is—how many times have you walked upstairs and forgotten why? Those characters will disappear immediately the first digit is entered but will return if all digits are deleted.

6.3.5   COMPLETION OF DATA ENTRY

With the tool design shown in Figure 6.4 the user can easily check both carb and time values before touching the OK button, whereupon (1) the interface fluidly returns to show the Food region (Figure 6.2); (2) the values of carb and time are recorded; and (3) the chosen carb value appears within the carb icon as shown in Figure 6.3. Any changes of mind are easily accommodated by repeating the procedure described above.

6.3.6   COMMONALITY

The design decisions discussed above apply without modification to the entry of protein, fat , and drink values and will therefore not be discussed. The corresponding signifiers can be seen in the view of the Food region presented in Figure 6.2, and also on the second page of Appendix 1.

6.3.7   CHANGE OF MIND

At any point a user must be able to change their mind without penalty, confusion, or inconvenience and without any earlier use of the tool being recorded. The standard and familiar means of supporting such a decision is to provide, usually at the top right of a tool, a black cross, as included in Figure 6.4. For consistency, the same approach is employed throughout the book for all tools (but see Advice, Chapter 12).

6.4   FURTHER DESIGN ISSUES

If the user chooses not to enter a time, a default value of the time at data entry will be recorded. And if no entry is made for carb value for some reason, a default value of zero will be assumed and, following acceptance indicated by a touch on the OK button (again, a different shape and colour from the other buttons), that zero value will be shown in the carb icon (Figure 6.2). As much as possible, I try not to have "alert" windows popping up in view of their distraction: the time needed to respond to them and the disturbance to the user's mental model are such that I therefore try to minimise (even eliminate) the number of "pop-up" so-called "helpful" messages that are so common and which can interrupt the flow of an entry, causing cognitive confusion and irritation. Even urgent advice can be handled without a pop-up window—that is precisely what the Advice region is for (see Chapter 12).

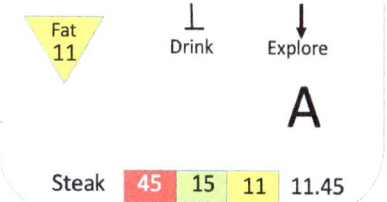

Figure 6.5: When the character "A" is touched a keyboard appears, allowing the entry of a name (e.g., "Steak" as in Figure 6.2)

Interest was expressed in entering the name of the planned meal and its display in the bar shown in Figure 6.2. That can be facilitated by using the single letter "A" (Figure 6.5) to signify that a touch will lead to the appearance of a keyboard.

The last decision to be made concerns the location of the carb icon in the Food region. Clinical advice indicated that the most important item of contextual information about Food that should be visible when any of the other three regions is "in front" is the carb value. That is why we have arranged for the carb icon to be located in the "always visible" area of the Food region (Figure 6.6).

6.4.1   DESIGN FLEXIBILITY

I make no apology for repeating an earlier discussion about design flexibility, particularly because we have made detailed design decisions about the Food region in this chapter. Such flexibility exists for the size of the personal regions in two respects (Figure 6.7). First, region content may vary considerably from one region to another, suggesting that region size (and maybe shape?) might be adjusted accordingly. Second, there may be advantage in temporarily expanding the area of the region currently at the front of the stack.

6.4.2   THE "BIG PICTURE"

At this point I suggest that the reader will benefit by turning to the second page of Appendix 1 and looking at the first horizontal entry. Why? First, because it concisely summarises the decision made in this chapter concerning the relation between a region (here, Food), a signifier (here, the carb icon) and the corresponding tool. But second, and of considerable importance, we see that the same "region/signifier/tool" scheme applies to other Food affordances—for the first four the tools are, in fact, identical. And if that page is turned, we see that the region/signifier/tool scheme applies to the other three personal regions. It could justifiably be argued that the consistency exhibited in those two pages goes some way toward benefitting the user's mental model and, in the present context, prepares the way for our consideration of the remaining personal affordances.

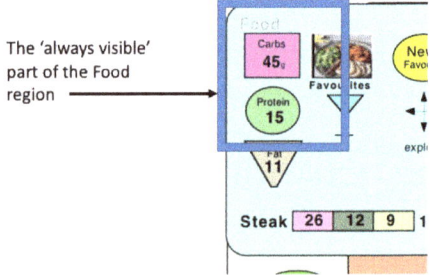

Figure 6.6: Icons and their associated values located in the "always visible" part of the Food region provide useful context, whatever the ordering of the other regions in the stack. The rectangle shown for explanation purposes is not visible to a user.

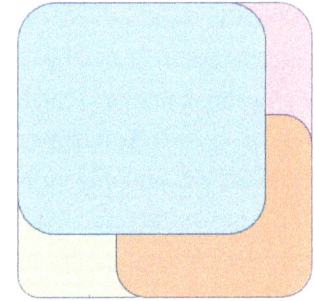

Figure 6.7: Illustrating flexibility in the choice of region size.

6.5   LEAF AFFORDANCES, PORTALS, AND TOOLS

In Chapter 3, we introduced a representation (Figure 6.8) of the major affordances required to be available in the proposed app. Extension to include sub-affordances like the one (carb) discussed above is shown in Figure 6.9 for the Food affordance. There is no change needed to the notation: a black ring denotes groups of affordances that are mutually exclusive in use and passes through their superordinate affordance.

Some affordances, e.g., "carb" have no sub-affordances, but instead lead directly to a tool such as that shown in Figure 6.3). Because the carb affordance has no sub- affordances it is reminiscent of a leaf node in a node-link diagram, so we shall refer to it as a **Leaf Affordance**.

The need to define a leaf affordance arises from the fact that, while the major Food affordance provides a portal to the carb sub-affordance, it is convenient to regard the carb sub-affordance as providing, not a portal to further sub- affordances, but a tool, in this case for the entry of carb value and time. We choose to represent the leaf affordance and its associated tool, as shown in Figure 6.10 by a grey circle containing a miniature representation of the tool shown in Figure 6.3. The grey circle emphasises that "carb" is not a portal, and contains no hierarchy involving exclusive multiple affordance use.

It could be argued, and correctly so, that the carb sub-affordance *does* have its own sub- affordances. Thus, the interface shown in Figure 6.3 affords the separate entry of all 10 individual digits and the use of the delete function and bistable clock, together representable by 12 individual sub-affordances on a black ring. But the affordance representation of the entire app would then become immensely extensive and of little extra value to the interface designer or anyone else. Thus, for purely pragmatic reasons, we speak of a single tool rather than a collection of individual affordances. It is a question of granularity of representation.

6.6   THE DIARY BRIEFLY REVISITED

In Chapter 5, we extended the functionality of the Diary by enabling temporal data to be presented in either the Bifocal or History mode. Both those affordances are leaf affordances and are mutually exclusive in use, so the affordance representation of the Diary now appears as in Figure 6.11.

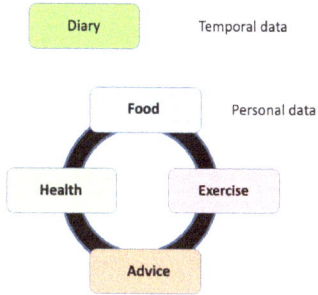

Figure 6.8: The major affordances to be offered.

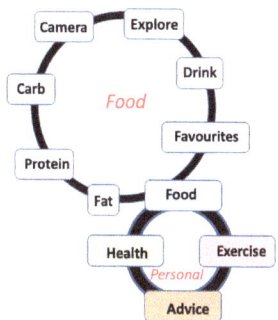

Figure 6.9: Showing the seven sub- ordinate affordances associated with the Food affordance.

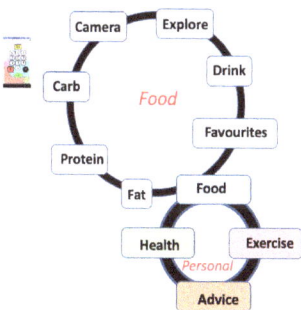

Figure 6.10: Incorporation, in the affordance representation, of the tool provided for the entry of carb value and time of consumption.

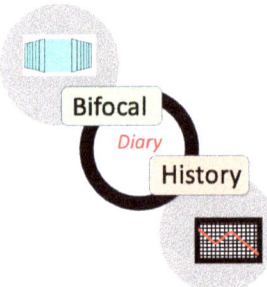

Figure 6.11: The affordance representation of the Diary. One of its two modes is always visible toward the top of the smartphone.

CHAPTER 7

Explore

It is extremely helpful if a user can manually and smoothly vary a carb value and immediately see the effect on predicted blood glucose level over the next hour or so, together with a recommended insulin dose. Such an activity is called 'dynamic exploration' and is made possible by our ability, based on machine learning, to make such a prediction.

In the previous chapter, we have seen how our app can support a user's entry of carb, protein, fat, and drink values, in each case employing a very simple region/signifier/tool scheme. In this chapter, we extend our consideration to another of the Food affordances, that of Dynamic Exploration, and find that the scheme is equally apposite, as illustrated in Figure 7.1. The Explore icon can be found in the Food region shown in Figure 7.2.

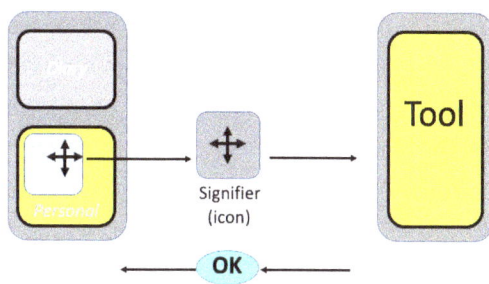

Figure 7.1: The region/signifier/tool scheme permitting dynamic exploration.

7.1 DYNAMIC EXPLORATION

In many situations people welcome the ability to ask "What if?" questions. In the context of our app, an example would be "What would happen to my blood glucose level in the next hour or so if I eat this particular meal?" While a specific answer to that question would certainly be useful, much more helpful would be the ability of a user to manually and smoothly vary a carb value and, immediately, see how their blood glucose level is predicted to vary over the next hour or so. Such an action is referred to as dynamic exploration.[4]

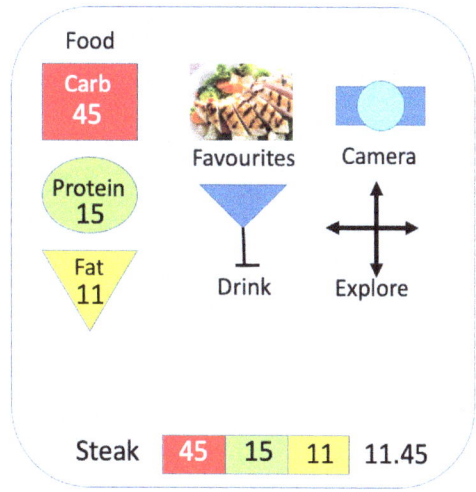

Figure 7.2: The Food region containing the Explore icon (at right).

---

[4] Dynamic exploration has wide relevance, and is made possible if a simple mathematical model or appropriate and readily accessible data is available. Spence (2014) provides a number of examples.

7.2   THE TOOL

The scheme to support dynamic exploration is shown in outline in Figure 7.1. A touch on the Explore icon in the Food region (Figure 7.2) causes the tool shown in Figure 7.3 to appear. The horizontal slider allows a carb value to be smoothly varied by moving a finger along that slider. Immediately, with essentially no delay, the corresponding carb value is indicated within the yellow box labeled "carb" (here, 20 grams). Within less than a second,[5] and in the black box (upper-right in Figure 7.3) the user will see not only the recorded blood glucose level up to the present moment but also, as indicated by a full line extending over the next hour or so, the predicted variation of glucose level for the current carb value, here 20. Dotted lines indicate confidence levels in the prediction. At the same time, a recommended insulin dose is shown in the yellow box just above the vertical slider. The appearance of that slider and its yellow button indicates that the insulin dose itself can also be manually varied. When it is, the predicted glucose level for a fixed carb value will again be shown in the diary segment.

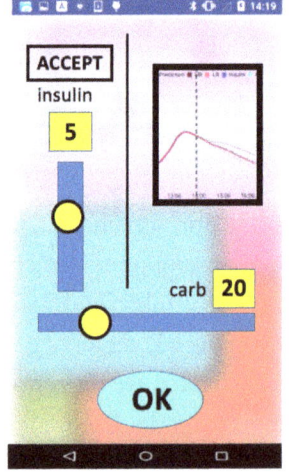

Figure 7.3: The Explore tool. Smooth manual variation of the carb value leads essentially immediately to a corresponding prediction of blood glucose level.

As indicated by the scheme shown in Figure 7.1, a return to the Food region of Figure 7.2 is initiated by a touch on the OK button, but can be preceded by acceptance of the recommended insulin dose.

7.3   BLOOD GLUCOSE LEVEL PREDICTION

To predict blood glucose levels, the ARISES system exploits machine learning as described by Li et al. (2019) and Daniels et al. (2020).

---

[5] Research has shown (Goodman and Spence, 1978) that in dynamic exploration of the sort described here, a user can tolerate a delay of as much as a second in the appearance of a prediction without loss of any effectiveness of the exploration procedure.

CHAPTER 8

Favourites

We are all creatures of habit, well illustrated by an individual's collection of favourite meals. A user should therefore have easy access to such a collection, and a subordinate affordance "Favourites" provides such access. Design of the Favourites affordance has much in common with Exercise in view of the convenience of a personalised scrollable menu of favourite meals.

Most people are creatures of habit, particularly in their choice of meals, so it was decided that the app should provide ready access to meals considered by individual users to be their favourites. So, we located, within the Food region, an icon that would lead to an appropriate tool. That icon (Figure 8.1) is in the middle of the top row of the Food region. It was thought that the best design for that icon would be a photograph of a meal, together with the word "Favourites." That photograph could be fixed or could be replaced by a photo of the meal last selected within the tool.

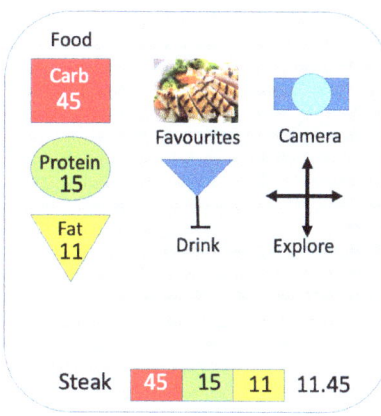

Figure 8.1: The Food region containing the Favourites icon.

8.1 A REJECTED TOOL

The first—and rejected—proposal for the tool was based on the mechanism shown in Figure 8.2. Some time ago I had proposed the "Carousel" tool (Spence and de Bruijn, 2002) as a means of rapidly examining the images within a folder. I thought that it could usefully provide a view of meal images moving sufficiently slowly that one could be chosen by a single touch, and proposed the tool shown in Figure 8.3. A touch on the red square would initiate a "carousel" presentation of favourite meals, allowing one to be chosen by touch. At this point, the focus group, as it usually did, provided immediate and valuable feedback: they firmly rejected the proposal. They pointed out that a *scrollable* collection of meal photographs would be best suited to the familiar action of scrolling using the thumb of the right hand while holding a smartphone.

Figure 8.2: The Carousel mode of Rapid Serial Visual Presentation. When interaction occurs, a stream of images emerges from one side of a folder, performs a roughly circular trajectory, and moves back into the folder.

Another good reason for rejecting the design of Figure 8.3 is the frustration of having to wait until an image of interest reappears. Following that welcomed feedback, the design of Figure 8.3 was discarded.

8.2   THE REVISED TOOL

The newly proposed tool is shown in Figure 8.4. As with many scrollable applications the movement of the collection of meal images is inertial, and has a "snap-to" property. The carb value is indicated for the meal whose image is identified by an arrow. The user can select the meal they wish to consume, and easily indicate if a smaller portion is needed. For convenience, the collection of meals is separated into the groups "breakfast", "meals", and "snacks". As with all other tools, a touch on "OK" confirms the choice and causes the Food region to fluidly reappear. Change of mind is supported by the "x" icon as in all other tools.

We note that the scheme proposed is consistent with the region/signifier/tool approach (Figure 8.5) employed for previously considered Food sub-affordances.

Figure 8.3: The (rejected) tool initially proposed to allow favourite meal choices.

Figure 8.4: The tool adopted for the Favourites affordance.

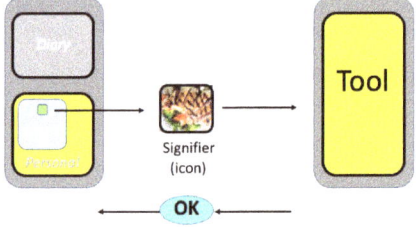

Figure 8.5: The region/signifier/tool scheme suited to the Favourites affordance.

8.3   PHOTOGRAPHED FAVOURITE

As will be discussed in the next chapter, the camera sub- affordance in the Food region allows the user to assign an already photographed meal to their collection of favourites.

.

CHAPTER 9

Photographs

As use of the app proceeds it is anticipated that a user will wish to make additions to their collection of favourite meals. They must therefore be able to photograph such a meal and separately arrange for its appearance in the scrollable collection of Favourites. With consistency with other sub-affordances in mind, we design the tool for such a situation.

As use of the app proceeds, it is anticipated that a user will wish to make additions to their collection of favourite meals. For that reason, we include within the Food region (Figure 9.1) a camera icon indicating an opportunity to place a photograph of an actual meal in the collection of favourite meals.

Smartphones already offer a powerful set of image capture functions, and they vary from one smartphone and operating system to another. We shall assume here that a meal picture has already been taken and identified as of interest. Then, selection of the photo icon (Figure 9.1, top right) will cause the tool shown in Figure 9.2 to appear. In this way, the region/signifier/tool scheme employed for other Food sub-affordances will again maintain consistency beneficial to a user's mental model. With that tool a new favourite meal image can be added to the existing collection, a value assigned to its carb value and the meal itself categorised as breakfast, meal, or snack.

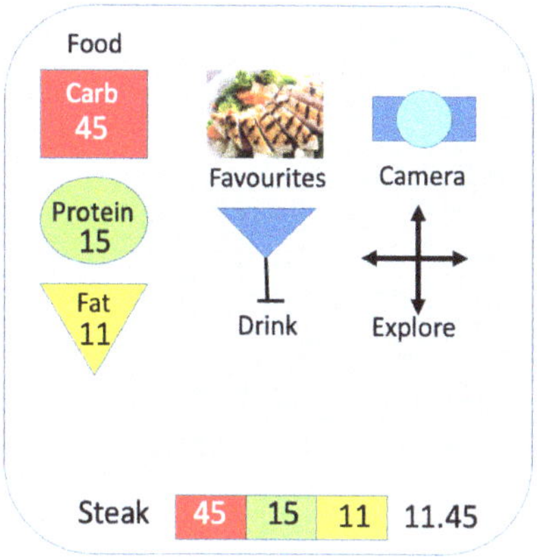

Figure 9.1: The Food region containing the photograph icon (top right).

9.1 COMMENT

The above concludes our consideration of the Food sub-affordances, all exploiting the region/signifier/tool scheme. That scheme will be found to be appropriate, in the next three chapters, to the remaining Personal regions, as a glance at Appendix 1 will confirm.

Figure 9.2: The tool supports the assignment of a carb value to a meal photo, and the inclusion of that dish in the Favourites collection.

CHAPTER 10

Exercise

Exercise is one of the four personal affordances. Here we design the tool that permits the entry of Exercise parameters such as type (e.g., squash), extent (e.g., 30 minutes), scheduled time (e.g., 4pm) and aerobic level.

Exercise is one of the four personal affordances. We assume for the moment that the Exercise region, previously not "at the front", has been touched to bring it to that position in the stack. The suggested appearance of the region is shown in Figure 10.1. As with all the personal regions the size of the Exercise region is sufficiently small that it is unsuitable to support the entry, by touch, of any detail. What the region can support is twofold.

First, it allows the user visually to review, in some detail, the exercise(s) planned for today. Second, when touched, it can lead to the appearance of a tool (Figure 10.2) that will allow the selection of new exercise(s).

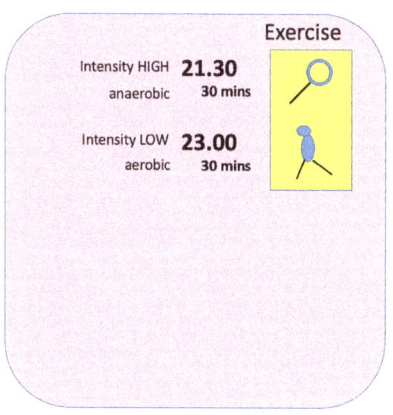

Figure 10.1: Appearance of the Exercise region.

10.1 THE TOOL

We see in Figure 10.2 the familiar keypad that can afford entry of exercise values (e.g., 1 km for Walk) as well as a scheduled time, the latter by use of the bistable clock icon. On the right is a (snap-to) scrollable menu of exercise types previously selected within *Settings* to reflect the user's principal interests. By default, the last choice of value (e.g., 1 km for Walk) appears against each available choice of type. Any item in that scrollable menu can be touched to indicate that it is intended for selection, whereupon its colour changes (green in Figure 10.2). In that state the scheduled time can be entered in addition to the value. When the user is happy with the selected exercise type, value, and scheduled time, a touch on OK will cause the new entry to be recorded as well as a fluid transition back to the updated region (Figure 10.1).

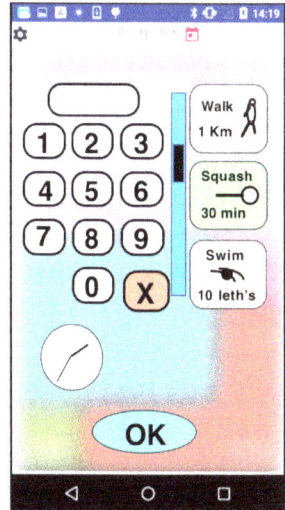

Figure 10.2: A first design of the Exercise tool.

10.2   THE SIGNIFIER

We have seen (Chapter 6) that, within the Food region, the carb icon with inset value (see Figure 10.3) serves two purposes. First, the user can observe the value previously assigned to the carb content and, second, it acts as a signifier to suggest that a touch will cause a transition to the relevant tool. On the assumption that consistency can be beneficial, we now look for an icon within which earlier choices of Exercise features can be viewed and which, when touched, leads to the appearance of the tool shown in Figure 10.2. To satisfy these requirements, I suggested that Exercise (type) choice should appear against the background of a yellow rectangle, the latter indicating a signifier sensitive to touch, and appearing in the "always visible" area of the Exercise region. Whether or not that decision—based on consistency—was useful may only be decided in usability trials.

Later discussions with the clinical lead led to the view that the ability to enter exercise intensity (three choices) and aerobic condition (two choices) would be beneficial for the remote calculation of recommended insulin doses. There is space (see Figure 10.5) for that additional feature, though there is now little spare room left for added features, a limitation that is discussed briefly below.

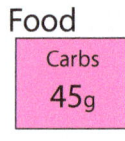

Figure 10.3: The carb icon within the Food region.

Figure 10.4: The Exercise icons within the Exercise region.

10.3   VISIBLE CONTEXT

An essential property of the Exercise region is that, following use of the tool, useful context information appears in the region's "always visible" area. That requirement influenced the design of Figure 10.1: the user will always see chosen exercise type(s) when other regions occlude the remainder of the Exercise region's content. Alternative content for the "always visible" region might be influenced by usability studies. Note that if the user needs a detailed reminder of exercise scheduled for today, a single touch on what can be seen of the Exercise region is all that is needed to bring it to "the front".

10.4   COMPLEXITY

A question that naturally arises, especially in view of the well-filled display of Figure 10.5, concerns a situation where more tool components are needed than can realistically be placed in the area available on the smartphone display. One simple approach is suggested by

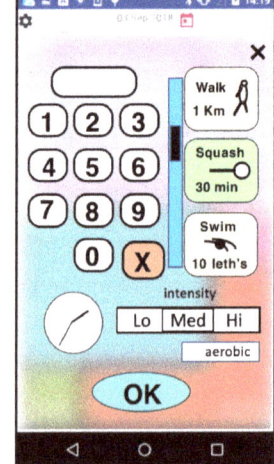

Figure 10.5: Support for the additional choice of intensity and aerobic state.

Figure 10.6. In the corresponding leaf affordance representation of Figure 10.7 the grey circle still contains the separate tools that are provided, but the black ring, as before, denotes their multiple exclusive use.

10.5   COMMENT

It could be argued that the yellow background shown in Figures 10.1 and 10.4 is justified only by a desire to maintain consistency (as demonstrated in Appendix 1). Experiments may or may not resolve that issue.

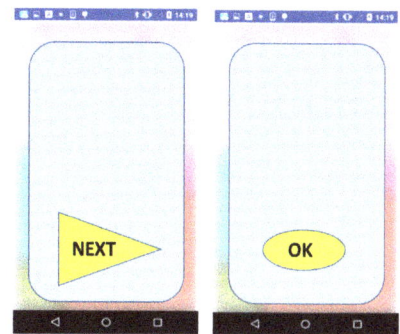

Figure 10.6: A tool separated into two presentations.

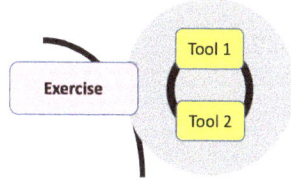

Figure 10.7: Representation of a Leaf affordance when more than one tool is needed. The black ring has the same meaning as before.

CHAPTER 11

Health

Health is one of the personal affordances. We design the appearance of its region and the tool that allows appropriate data to be entered. We also use the Health region to illustrate a simple way of determining what temporal data will be presented in the Diary, an approach that generalises to all personal regions.

Health is one of the four personal affordances, and its design is relatively straightforward, though we bear in mind the anticipated benefits that can accrue from ensuring its design is consistent with other personal affordances, as illustrated in Appendix 1.

11.1 A REJECTED DESIGN

A very early experimental design of the Health region, allowing variables to be chosen for display in the Diary, is shown in Figure 11.1. I considered it unacceptable, for three important reasons. First, its complexity could be confusing for a user.

Second, as with the other regions, the size of the Health region makes it an unacceptable place to locate many (over 20) icons that need touching with precision—that is what a tool is for.

Third, and most importantly, it ignores the very purpose of the Diary which is to allow the detailed examination of temporal data. A new design was essential and is described in this chapter.

11.2 THE NEW DESIGN

Clinical advice indicated that the Health region only requires the user to enter and view three items of data, one of which applies only to female users. They are Stress, Illness,

Figure 11.1: An early, but rejected, experimental design of the Health region.

Figure 11.2: Initial revised design of the Health region.

and Menses, each of which can be treated as bistable. As a result, a new design was proposed for the

Health region: its initial form is shown in Figure 11.2. As with the other three personal regions, the available space is sufficiently small that its function is just twofold: to allow the states of the three items of data to be reviewed with ease, and to provide access to the tool (Figure 11.3) that will allow the user to change the state of each of those three items. Consistently with the Exercise region design, the yellow area shown in Figure 11.2 supports the user's review of the current state of the Health data in the "always visible" area and indicates where a touch is required to cause the tool to appear.

11.3   GRAPHICAL PRESENTATION OF DATA

We introduce considerable simplification into the way in which temporal data is selected for view by including, in the Health region, a single graph icon (Figure 11.4) having two states. Its function is to control the Health data that is presented in the Diary. In one sense, therefore, it is a replacement for much of the rejected design of Figure 11.1. A single touch on the graph icon, automatically confirmed visually by highlighting, causes default health data to be displayed in the selected mode of theDiary. If one or more of the three Health items is then selected, that selection will be highlighted and the corresponding data will also be presented in the Diary. Deselecting the bistable graph icon automatically causes **deselection** of the plotting of all Health data.

11.4   GENERALIZATION

A distinct anticipated advantage of this proposed approach to the identification of data to be displayed in the Diary is that it generalises to all Personal regions. The graph icon is therefore included in all the regions, as shown in Appendix 1.

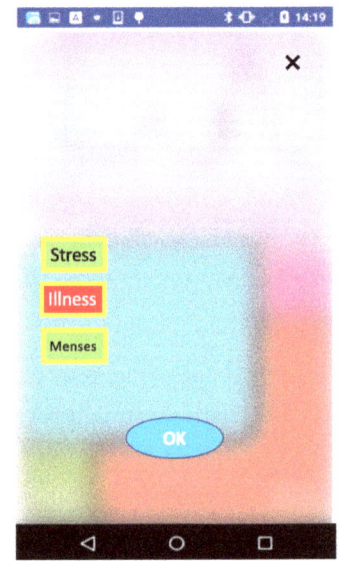

Figure 11.3: The tool associated with the Health region. A touch on any one of the three bistable buttons will change its state.

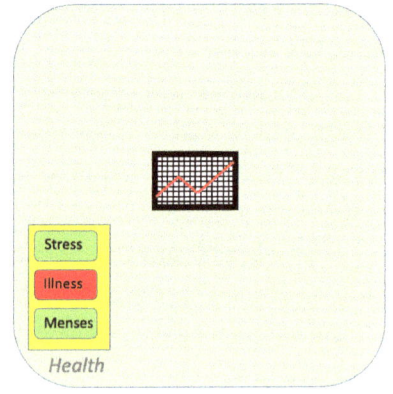

Figure 11.4: The graph icon controls the Health data shown in the Diary.

11.5   LEAF AFFORDANCE

Since the Health affordance has no sub-affordances and is therefore a leaf affordance, it is represented as shown in Figure 11.5, with the grey circle containing a miniature representation of the tool.

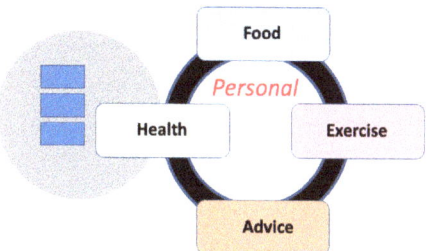

Figure 11.5: Health is a leaf affordance.

CHAPTER 12

Advice

One essential function of the app is to provide advice to the user. That advice may range from regular recommendations concerning insulin doses to alerts regarding blood glucose levels that require urgent acknowledgement and attention. Conventionally, alerts are often delivered by pop-up windows, but the rationale behind the approach adopted here is that the user's mental model would suffer from such an additional dialogue feature and that the correct location of advice is in the Advice region and its tool

12.1   HOW TO PROVIDE THE USER WITH ADVICE

Although Advice is one of the four personal affordances it differs from the other three in that it is the means by which the system provides advice to the user. Examples include recommendations regarding insulin doses and, in urgent situations, alerts. The latter might be triggered by the approach of the user's blood glucose level to the dangerous hyper- or hypo-glycaemic levels.

   One approach to the issue of alerts favors the use of "pop-up" windows. I hold the different, and perhaps controversial, view described below for two reasons. First, the very name "Advice" is associated with one of the four Personal regions, and so is already available to enable the system to advise the user. Second, if one is trying to enhance usability, one should minimise the number of features that the user would have to include in their mental model of the interface. Typically, a pop-up alert window is intended to suggest urgency, but I would argue that, instead, the sudden appearance of the Advice tool—possibly with accompanying sounds, graphics, and/or vibrations—is equally capable of attracting a user's attention.

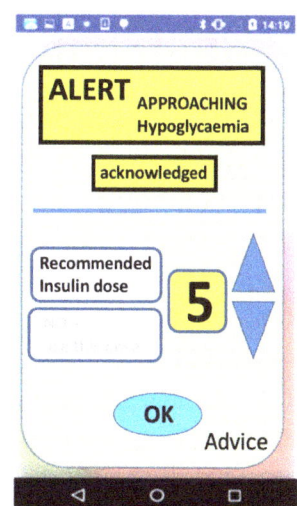

Figure 12.1: The Advice tool. It may appear automatically in the case of an alert, or at the request of the user after a touch on the Advice region when at the front of the stack.

12.2   AN ALERT

If the user needs to be alerted or advised, I suggest that the Advice tool—not its region—should immediately appear, and appropriate sounds and/or vibration should additionally be used to engage the user's attention. That tool is shown in Figure 12.1 but needs explanation. Alerts are located in the upper part of the tool and appear in a box. Underneath the box is a button containing the word "acknowledged": a touch on that button will tell the system that the user has seen the alert message. It may well be judged that, until that word is touched, all else in the tool is greyed out, including the OK button.

With all the other tools it is possible at any point for a user to change their mind without penalty, confusion, or inconvenience and without any earlier use of the tool being recorded, simply by touching the "x" symbol conventionally located at the top right of a tool. For the Advice affordance I have intentionally left this provision open (and hence absent from Figure 12.1) to allow a clinical decision to be made.

12.3   RECOMMENDED INSULIN DOSE

Now we examine the lower part of the tool. Figure 12.2 shows an enlarged example of its initial appearance. It is suggested that the user will assume, in the example shown, that an injection of five units of insulin is recommended and that a touch on OK will lead to that dose being administered. Consistent with other tools (see Appendix 1), the Advice region, discussed below, will then replace the tool.

The user will also notice from the other (greyed out) button that they have the option of declining that recommendation. A touch on that button will lead to three changes (see Figure 12.3): (1) the "NO" button will be highlighted; (2) concurrently, the 'recommended' button will be greyed out; and (3) two arrows will (fluidly) appear and can be used to increase or decrease the integer that denotes an insulin dose. The step from Figure 12.2 to 12.3 is reversible. When the user is satisfied with their choice of dose, a touch on OK will again cause the complete Advice region to replace the tool.

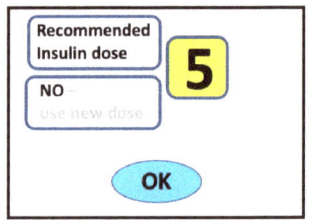

Figure 12.2: Initial appearance of the lower part of the Advice tool. A touch on OK indicates acceptance of the recommended insulin dose.

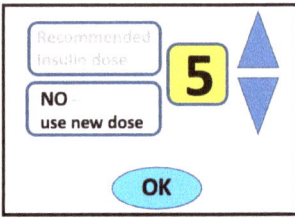

Figure 12.3: The result of the user selecting the "NO" option, which is then highlighted. Two arrows fluidly appear and allow the selection of an alternative insulin dose. A touch on OK then confirms that choice.

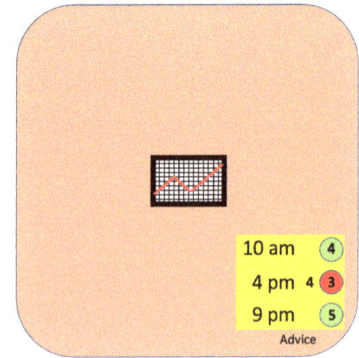

Figure 12.4: Showing, in the "always visible" area of the Advice region, a summary of the day's administration of insulin.

12.4   ADVICE REGION

The Advice region is shown in Figure 12.4. As far as possible, our aim is to design the Advice region to be consistent with other personal regions. The contextual information shown in Figure 12.4 may be the only content of the Advice region, although there is plenty of space to handle other clinical requirements. In the "always visible" area it lists the recommendations for insulin doses and the times they were made, together with the number of units administered. If the background to a number is red instead of green it is an indication that the units chosen by the user are different from those that were recommended, with the recommendation shown beside the circled number. It is expected that the "always visible" area of the region is sufficient to display a day's record. Consistent with the other Personal regions, Advice contains a graph icon (Figure 12.4) that allows variables such as advice and dose recommendations over time to be inspected within the Diary.

12.5   LEAF AFFORDANCE

Advice is a leaf affordance, and is represented as shown in Figure 12.5

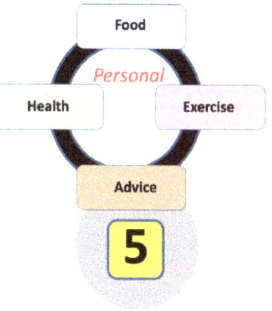

Figure 12.5: Advice is a leaf affordance.

CHAPTER 13

A Dialogue Check

The aim of an app is to support an effective and enjoyable dialogue. Experience has taught us, the hard way, that many undesirable features can lead to apps that do not fully achieve that aim. In this Chapter we check our design against some well-established guidelines, and add some other guidelines on which I personally have placed priority.

Whatever interaction takes place, the user of our app is engaged in a dialogue. Like any dialogue it may well be complex, but it is the duty of an app designer to ensure that it is as intuitive, effective and enjoyable as possible. Many guidelines exist that enable an app designer to assess and influence the acceptability of their design, and I have chosen to focus on the comprehensive set of eight guidelines set out by Elmqvist et al. (2011). Below, I copy each guideline (DG*) from their paper and in each case provide my own view as to the extent to which they are satisfied by the app we have designed. I have also presented, in Section 13.2, other guidelines on which I have placed priority.

13.1   CHECK OF GUIDELINES

1. **DG1: Use smooth animated transitions between states.** This requirement has been specified in this book and repeated many times in view of its importance. The coder has the responsibility to ensure implementation.

2. **DG2: Provide immediate visual feedback on interaction.** This requirement has also been specified in this book: it is the responsibility of the coder to ensure implementation.

3. **DG3: Minimise indirection in the interface. If possible, use direct manipulation so that interactive operations (e.g., filtering, selection, and details-on-demand) are integrated in the visual representation. In particular, avoid control panels that are separated from the *visualisation*.** The app contains no control panels. Through use of direct manipulation there is a very close relationship between an interaction and the corresponding visual change.

4. **DG4: Integrate user interface components in the visual representation.** My interpretation of this guideline in the present context is that it is close to DG3. A detailed check of the designed interface suggests that the spirit of DG4 is fully recognised.

5. **DG5: Reward interaction.** In Elmqvist et al. (2011), this guideline referred to the activity of data *visualisation* and is not particularly relevant to our design challenge except that, if the app behaves as the user expects, that in itself might be considered a reward. In that sense, I suggest that DG5 may be satisfied.

6. **DG6: Ensure that interaction never "ends". The user should never reach a dead end where they can no longer proceed.** Our app design satisfies this guideline.

7. **DG7: Reinforce a clear conceptual model. The user should always have a clear idea of the state of the (visualisation) and all interactions should be designed to reinforce this model. Operations should be reversible, allowing the user to return to a previous state.** I would argue that the need for a clear conceptual model is satisfied by many features of our app. A primary feature in this respect is our choice of the two metaphors introduced in Chapter 4. Another confidently suspected beneficial feature is the continuous visibility of major components of the interface, albeit at times necessarily partial and/or inconspicuous. A very minor exception arises from the two mutually exclusive modes of the Diary. Most operations are reversible. Those few that are not include, for example, the movement of a region to the front of the Stack: subsequently it would not immediately be obvious how to restore the original ordering which, of course, is of little consequence.

8. **DG8: Avoid explicit mode changes. Instead of introducing different modes, integrate all operations in the same mode. This includes avoiding both drastic visual changes and drastic interaction modality changes. Mode changes may break the user's flow.** I think we have satisfied this guideline to the extent possible. The exception is the Diary which can exist in one of two modes. My own view is that this will not cause undue difficulty, though a usability study may of course prove me wrong.

13.2   ADDED GUIDELINES

In addition to the guidelines established by Elmqvist et al. there are many other guidelines available in the literature, many of which I have adopted without citation. Below, I have presented three guidelines on which I have personally placed great importance and which, I think, will be major influences on the usability of our—or any other—app. Unsurprisingly, overlap with some of Elmqvist et al.'s guidelines may be observed.

1. **Visual Context 1: To minimise navigational difficulties, provide visual context indicating available transitions.** One major advantage (for the others, see below) of

the Stack metaphor is that available "next step" destinations are in full view. Such a benefit has long been known, of course, and incorporated in conventional tags.

2. **Visual Context 2: To support a user's awareness of personal conditions that together affect their wellbeing, provide as much relevant information as possible.** This guideline prompted the provision of relevant information in the "always visible" areas of all four regions. Nevertheless, it identifies a trade-off that requires a designer's attention: namely, that the region "in front" can be expanded if that helps to support interaction, but with a corresponding reduction in the "always visible" areas of the three partially masked regions.

3. **Visual Context 3: Ensure that the user's attention is not unduly attracted by irrelevant visible material.** This guideline is the basis for our decision, for example, to employ a defocused background for the tools available for data entry. We have used the term "inconspicuous visible context" to refer to this approach.

CHAPTER   14

Conclusions

One can only draw subjective conclusions about the usability of an app that, principally for pandemic considerations, has not been evaluated. Nevertheless, for debate, five subjective claims are presented.

14.1   USABILITY

One conclusion that *cannot* be drawn about the app's design concerns its usability: only an independent study can provide appropriate evidence. Unfortunately, restrictions imposed by the Covid-19 virus have curtailed many activities, and a clinical trial of the ARISES app is one of them. Even planned focused and limited studies were impossible to arrange. The conclusions to be drawn are therefore subjective, and will in many respects repeat, though briefly, various comments about the benefits I anticipated.

14.2   ANTICIPATED BENEFITS OF THE APP

My necessarily subjective view of the app's effectiveness has been voiced at various stages of its development and will not be repeated in that form. Instead, I'll list just five briefly expressed subjective claims for the final design in the context of its usability. Each item can be—indeed should be—subjected to comprehensive critique and, best of all, experimental study.

- **Claim 1.** The two metaphors adopted (Bifocal display and the Stack of regions) support an enhanced mental model of the app's behavior.

- **Claim 2.** The provision of visible structural context minimises navigational difficulty.

- **Claim 3.** The visible context provided by the "always visible" areas of the personal regions provides the user with a more comprehensive awareness of their wellbeing.

- **Claim 4.** Inconspicuous visible context will minimise the occurrence of unexpected navigational problems.

- **Claim 5.** The consistency exhibited by the "region-signifier-tool" scheme for personal regions will reduce the cognitive load on the user.

It would be surprising if these (subjective) claims were not found to be controversial to some degree. If so, they should stimulate productive discussion.

14.3   THE AFFORDANCE CONCEPT

The concept of affordance has influenced my thoughts about interface design. On reflection, the concept would appear to benefit both collaboration within a team and the experimental study of usability, the latter as suggested in Appendix 2. Again, these are subjective views and therefore unproven, though certainly worth debate.

CHAPTER 15

Reflections on Affordance and Design

In this brief chapter I reflect upon the concept of affordance and the way it has influenced the progress of design, and suggest how the affordance representation might beneficially be used in a physical design environment.

15.1 COMMUNICATION FOR COLLABORATION

I first became more than a little interested in the concept of affordance well into the process of app design. At first (Spence and Redmond, 2020a) it seemed that the affordance representation could be valuable in studies of usability: the record of a user's sequence of interactions could easily be transferred to the affordance diagram (Figure 15.1, also Appendix 2) and provide some insight into user behavior, especially including circumstances where part of a dialogue was unexpected. But later (Spence and Redmond, 2020b), on reviewing the design of the app, it became obvious that one potential advantage of the affordance representation was its potential as a notation, offering a basis for communication between the many actors involved in the realization of a working app—the visionary, the medical lead, the interaction designer, the usability expert, and even the coder. That realization led to further thoughts concerning design.

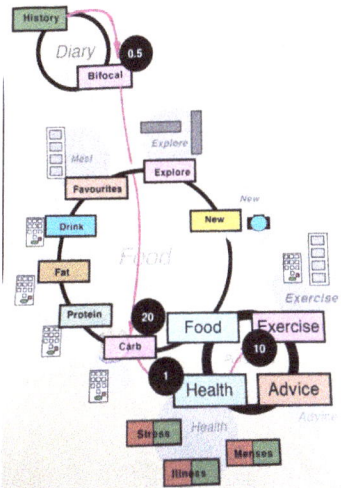

Figure 15.1: The app's affordance representation annotated with recorded interactions to facilitate a usability study.

15.2 A VISION

Thoughts about collaboration triggered a reminder of something that happened to me in 1995. That year I interviewed, independently, 12 eminent engineers to elicit their visions of engineering design 25 years later, in the year 2020 (I must admit that the expression 20/20 vision did come to mind!) (Spence, 1995). One vision that was put forward concerned the way in which the advantage of collaboration could be encouraged. An illustration (Figure 15.2) of that vision was set in an imaginary small company involved in the design and manufacture of a device of nano proportions that could be injected into a human being and transmit useful clinical data to an external receiver. It was envisioned that a scaled-up model of the device would usefully be placed in a coffee room and thereby enhance any collaboration that might otherwise take place.

15.3   DESIGN ENVIRONMENT

That vision is not too dissimilar from a collaborative work environment (Figure 15.3) that I encountered during a workshop I presented in Portugal. That leads to my proposal (Figure 15.4) that a large wall presentation of an affordance representation might well provide effective support to the many collaborators who provide the wide range of expertise required for an app's development. Annotations (removable) and ideas contained on Post-it notes would be a valuable feature of the dialogue supported by the design tool and enable the representation to "grow" over time.

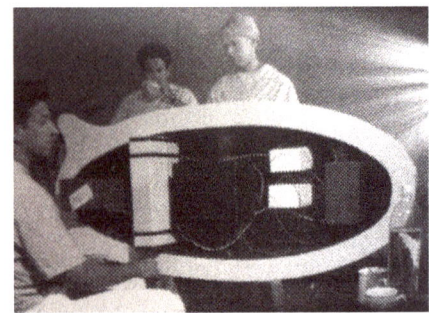

Figure 15.2: A 1995 vision of a collaborative design scenario.

Figure 15.3: Collaborative design at the University of Madeira, Portugal.

Figure 15.4: Possible use of an affordance representation in a design environment.

CHAPTER 16

Colleagues

I did not work alone on the interface design: I drew inspiration and advice from a variety of people, some concerned directly with the project, some not. They are not included as coauthors for the simple reason that this book contains an expression of my own very personal views about interface design.

I did not work alone on the interface design: I drew inspiration and advice from many people, some concerned directly with the project, some not. And others were members of the many focus group meetings that provided such valuable feedback. They are not included as coauthors for the simple reason that this book is an expression of my own very personal views about interface design. The list that follows is a rather inadequate way of saying "Thanks".

Pantelis Georgiou is the visionary behind the ARISES project and was there to provide guidance and encouragement when needed. He is Reader in Biomedical Electronics in the Department of Electrical and Electronic Engineering of Imperial College.

Chukwuma Oduku is the clinical lead on the ARISES project and his knowledge of Type 1 diabetes was crucial to the app's development. Our many meetings, as well as the focus meetings he chaired, were pivotal to my proposals for the interface.

Nick Oliver is Wynn Professor of human metabolism and is a consultant in diabetes and endocrinology at Imperial College Healthcare NHS Trust. A very early discussion with Nick encouraged me to propose the inclusion of the Explore affordance in the app.

Leah Redmond is an undergraduate (now third year) in the Department of Bioengineering of Imperial College. It was during a short collaboration during the summer of her first year that the value of an affordance representation of the app occurred to us.

Mark Apperley is Professor of Computer Science at the University of Waikato in New Zealand. Since 1971 we have been collaborating in research in the general area of Human–Computer Interaction. I have learned a lot, and applied it, as a result of that long collaboration. Mark specifically suggested that the Diary should have a History mode, something that was immediately recognised as blindingly obvious—but then, many good ideas are like that.

Jugnee Navada organized and recorded the focus group meetings whose outcomes were so valuable to the ongoing discussion of interface design.

Ken Li joined me in many meetings associated with the interface design and provided an excellent "sounding board" for my proposals.

Oscar de Bruijn is a colleague of many years' standing whose comments on my interface work are always so valuable.

Taiyu Zhu provided valuable advice regarding the implementation of the ARISES interface.

Interaction Consistency

In what follows, you will find a summary of the transitions between regions and tools, as well as between tools for the Diary. The purpose is two-fold:

> [1] to highlight the consistency exhibited by the interface design, a property that will hopefully simplify the mental model of the user; and

> [2] serve as a reference to the major design decisions taken. To support such reference the figure numbers have been identified within black circles.

Region (in front of stack)	Signifier (one touch causes transition)	Tools (with figure references)

Region (in front of stack)	Signifier (one touch causes transition)	Tools (with figure references)

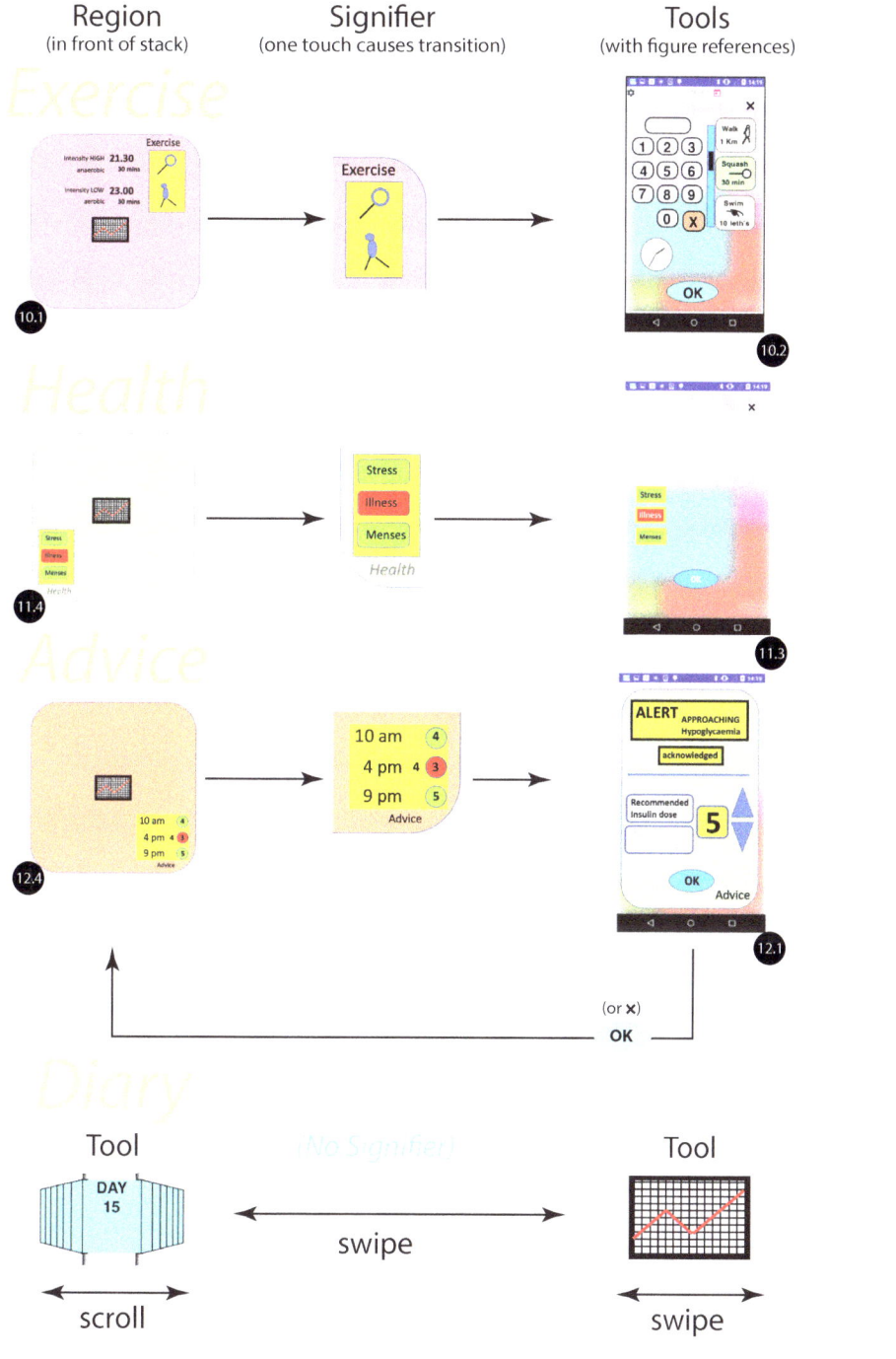

Exercise

Health

Advice

Diary

Tool	(No Signifier)	Tool

swipe

scroll

swipe

APPENDIX 2

A Novel Usability Tool

During use of the app's affordance representation, there emerged a proposal for a new approach to studying the app's usability.

Our consideration of affordances during the design of the app led unexpectantly to a proposal (Spence and Redmond, 2020a) for investigating an app's usability.

It was proposed that if a user's interaction with the app was recorded, the corresponding trajectory (i.e., of touches, scrolls, and swipes) could usefully be sketched on the app's affordance representation, as illustrated by the pink line in both Figure 15.1 and Figure A2.1. That line shows the following action of the user. First, they change the existing Diary mode (History) to Bifocal before deploying the carb affordance. Then, after unintentionally selecting the health affordance they quickly select the Exercise affordance. The numbers in the black circles record what we might call "dwell times": 0.5 sec for the Diary's Bifocal mode; 20 sec to enter a carb value and associated time; 1 sec to recognise an unintended selection of Health; and finally, 10 sec to select an Exercise.

The perceived benefit of the representation shown in Figure A2.1 is that a usability expert (though in practice probably the interface designer initially) can gain an impression of how user performance varies from that which was expected. That impression may well contribute to plans for design change. That benefit might well be enhanced if trajectories for several users asked to perform the same task were superimposed.

No apology is offered for the pragmatic approach just described. Its absence of elegance, mathematical grounding, and statistical considerations is intentional, and does not preclude any more formal investigations that might occur later.

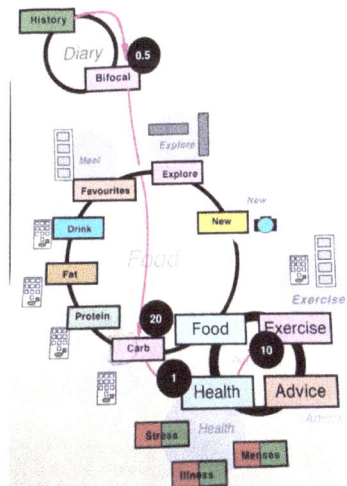

Figure A.2.1: Use of the app's affordance representation to investigate usability. The sketched pink line shows a record of a user's interactions, and the numbers in black circles indicate "dwell times" in seconds.

References

Cheyette, C. and Balolia, Y. (2016). *Carb and Calorie Counter*. Chello Publishing Company. 23

Daniels, J., Zhu, T, Li, K., Uduku, C., Spence, R., Herrero, P., Oliver, N., and Georgiou, P. (2020). ARISES: an advanced clinical decision support platform for T1D Management. *Diabetes Technology & Therapeutics*, 22:A57–A57. 4, 30

Dourish, P. (2001). *Where the Action Is: The Foundations of Embodied Interaction*. Cambridge, MA: MIT Press. DOI: 10.7551/mitpress/7221.001.0001. 3, 14

Elmqvist, N., van der Moere, A., Jetter, H-C., Cernea, D., Reiterer, H., and Jankun-Kelly, T.J. (2011). Fluid interaction for information visualization. *Information Visualization*, 10(4):327–340. DOI: 10.1177/1473871611413180. xxiii, 19, 20, 49, 50

Findlay, J.M. and Gilchrist, I.D. (2003). *Active Vision: The Psychology of Looking and Seeing*. Oxford: Oxford University Press. DOI: 10.1093/acprof:oso/9780198524793.001.0001. 19

Gibson, J.J. (1979). *The Ecological Approach to Visual Perception*. Psychology Press. 3

Goodman, T.J. and Spence, R. (1978). The effect of system response time on interactive computer aided problem solving. ACM, *SIGGRAPH Conference Proceedings*, pp.100–104. DOI: 10.1145/965139.807378. 30

Healey, C. (2007). Perception in Visualization, http://www.csc.ncsu.edu/faculty/healey/PP. 19

Hinton, A. (2015). *Understanding Context*. Beijing: O'Reilly. 3, 14

Kaptelinin, V. (2013). Affordances and design. *Design Studies*, 34(3):285–301. DOI: 10.1016/j.destud.2012.11.005. 3

Li, K., Daniels, J., Liu, C., Herrero, P., and Georgiou, P. (2019). Convolutional recurrent neural networks for glucose prediction. *IEEE Journal of Biomedical and Health Informatics*, 24:603–613. DOI: 10.1109/JBHI.2019.2908488. 30

Nielsen, J. (2010). Mental models. http://www.nngroup.com/articles/mental-models. October 17, 2010. 9

Norman, D. (1988). *The Design of Everyday Things*. Basic Books. 3

Rensink, R.A., O'Regan, J.K., and Clark, J.J. (1997). To see or not to see: The need for attention to perceive changes in scenes. *Psychological Science*, 8(5):368–373. DOI: 10.1111/j.1467-9280.1997.tb00427.x. 19

Robertson, G.G., Mackinlay, J.D., and Card, S.K. (1991). Cone trees: Animated 3D visu-
    alizations of hierarchical information. ACM, *Proceedings of CHI'91*, pp. 189–194.
    DOI:10.1145/108844.108883. 20

Simons, D.J. and Lewin, D.T. (1997). Change blindness. *Trends in Cognitive Sciences* 1(7): 261–267.
    DOI: 10.1016/S1364-6613(97)01080-2. 19

Spence, R. (2014). *Information Visualization: An Introduction*, 3rd edition. Springer. DOI:
    10.1007/978-3-319-07341-5_1. 29

Spence, R. (1995). Visions of design. *Journal of Engineering Design*, 6(25–137). DOI:
    10.1080/09544829508907908. 55

Spence, R. and Redmond, L. (2020a). Circles of affordance. Short paper, AVI conference. DOI:
    10.1145/3399715.3399719. 55, 63

Spence, R. and Redmond, L. (2020b). A new notation for interactive systems. Poster, AVI confer-
    ence. DOI: 10.1145/3399715.3399940. 55

Spence, R. and Apperley, M.D. (1982). Data base navigation: an office environment for the professional.
    Behaviour and Information Technology, 1(1):43–54. DOI: 10.1080/01449298208914435.
    10

Spence, R. and de Bruijn, O. (2002). Image presentation and control. US Patent application. 31

Spence, R., Li, K., Uduku, C., Zhu, T., Redmond, L., Herrero, P. Oliver, N., and Georgiou, P. (2020).
    A novel hand-held interface supporting the self-management of Type-1 Diabetes. *Proc.
    13th Int. Conf. on Advanced Technologies and Treatments for Diabetes*, pp. A58–A58. 1

Author Biography

Bob Spence is Professor Emeritus of Information Engineering and Senior Research Investigator at Imperial College London.

Following a two-year postdoctoral experience at General Dynamics/Electronics in the United States, he came to Imperial College in 1962, first as a Lecturer with interests in circuit theory. He is, in fact, a co-author of the classic *Tellegen's Theorem* (MIT Press, 1970) which powerfully generalizes the original form of that unusual theorem. But it was the emergence of interactive computer graphics in the 1960s that triggered interest in the actual user of that technology and, hence, in Human–Computer Interaction, a field of life-long interest and of which he is a pioneer. In this context, Bob has been the co-architect of a number of tools: first MINNIE (a powerful graphic front-end for circuit designers), then a CAD system exploiting Artificial Intelligence and, during the last two years, the interface of a hand-held device supporting the self-management of Type-1 diabetes, the topic of this book. At the same time, Bob co-invented the fish-eye lens as well as a collection of Exploration tools and a new notation for interactive systems and has collaborated extensively on research into Rapid Serial Visual Presentation.

His undergraduate teaching, which he enjoys, reflects his long-standing interest in Information Visualization, which has led to three editions of his book *Information Visualization*. He regularly presents workshops on Design for Information Visualization in Italy, Portugal, The Netherlands and elsewhere.

Bob's research has led to the award of three higher doctorates: on Circuit Theory (University of London, 1984), Interaction Design (Royal College of Art, 1995), and Interactive Visual Artefacts (Imperial College, 2018). He was elected to the Fellowship of the Royal Academy of Engineering, the equivalent of the U.S. National Academy of Engineering, in 1991.

Lightning Source UK Ltd.
Milton Keynes UK
UKHW052049200621
385843UK00001B/6